Postharvest Extension and Capacity Building for the Developing World

World Food Preservation Center Book Series

Series Editor

Charles L. Wilson

Postharvest Extension and Capacity Building for the Developing World

Majeed Mohammed and Vijay Yadav Tokala

For more information about this series please visit:
http://worldfoodpreservationcenter.com/crc-press.html

Postharvest Extension and Capacity Building for the Developing World

Edited by
Majeed Mohammed
Vijay Yadav Tokala

CRC Press
Taylor & Francis Group
Boca Raton London New York

CRC Press is an imprint of the
Taylor & Francis Group, an **informa** business

CRC Press
Taylor & Francis Group
6000 Broken Sound Parkway NW, Suite 300
Boca Raton, FL 33487-2742

First issued in paperback 2021

© 2019 by Taylor & Francis Group, LLC
CRC Press is an imprint of Taylor & Francis Group, an Informa business

No claim to original U.S. Government works

ISBN 13: 978-1-03-223751-0 (pbk)
ISBN 13: 978-1-138-06928-2 (hbk)

DOI: 10.1201/9781315115771

Library of Congress Cataloging-in-Publication Data

Names: Mohammed, M. (Majeed), editor.
Title: Postharvest extension and capacity building for the developing world / [edited by] Majeed Mohammed, Vijay Yadav Tokala.
Description: Boca Raton, Florida : CRC Press, [2019] | Series: World Food Preservation Center book series | Includes bibliographical references.
Identifiers: LCCN 2018030138| ISBN 9781138069282 (hardback : alk. paper) | ISBN 9781315115771 (ebook)
Subjects: LCSH: Crops--Postharvest technology--Developing countries.
Classification: LCC SB129 .P659 2019 | DDC 630.9172/4--dc23
LC record available at https://lccn.loc.gov/2018030138

Visit the Taylor & Francis Web site at
http://www.taylorandfrancis.com

and the CRC Press Web site at
http://www.crcpress.com

Contents

SECTION I FLW Assessment Methods

SECTION II Capacity Building

Foreword

As a compilation of the current practices and examples of models for improved postharvest training and capacity building for developing areas, this book provides a glimpse into a fast-changing area of agricultural extension research and postharvest applications. The concept for this book was presented to the board of directors of The Postharvest Education Foundation (PEF) by Dr. Charles L. Wilson of the World Food Preservation Center® LLC in 2016, and we agreed to reach out to our fellow postharvest extension and training professionals and assist him to identify the key topics and develop the chapter outlines. Two of our PEF board members took on the enormous job of editing this book project. Several of our board members contributed chapters on their recent extension work, including Dr. Diane Barrett (on creating postharvest videos) and Dr. Deirdre Holcroft (on postharvest e-learning programs). I was pleased to be able to contribute to a few chapters on topics falling within my areas of expertise, working with close colleagues Dr. Majeed Mohammed of the University of the West Indies, Dr. Jorge Fonseca of the Food and Agriculture Organization of the United Nations, Dr. Farbod Youssefi of the World Bank, and Dr. Devon Zagory, one of the founding board members of PEF.

I am also very pleased to see chapters based on successful postharvest extension and value addition projects undertaken in Egypt, India, and Tajikistan, as well as postharvest extension models that I have designed for development projects in India (chapter by Dr. B.V.C. Mahajan at Punjab Agricultural University) and Tanzania (chapter by Dr. Ngoni Nenguwo and colleagues at the World Vegetable Center in Arusha).

The book covers a wide range of topics, from Section I on postharvest loss assessment methods to Section IV on extension models. Each section of the book provides currently available details, but the field is growing exponentially as new projects begin and older projects and programs are evaluated and improved. Loss assessment methods are being improved and expanded, as various organizations are providing training on the methods they are championing. Postharvest capacity building is underway on a global scale via programs being managed by the Food and Agriculture Organization, the World Food Preservation Center, and various universities and international non-governmental organizations (NGOs). Postharvest training methods being used in the field include mentor guided e-learning programs, Massive Open Online Courses (MOOCs), animated videos, Information and Communication Technologies (ICTs), live demonstrations, and more. Comprehensive postharvest extension platforms are being launched in many countries, incorporating best practices in extension education, postharvest outreach, and support services.

The importance of all these diverse examples is that they are intended to serve as models for future postharvest extension work. I hope they will be adopted, adapted, and further improved by the readers of this book as they continue to work on reducing food losses and waste in their home countries.

Lisa Kitinoja
The Postharvest Education Foundation

Preface

It is estimated that around 1.3 billion metric tons per year of food produced for human consumption, which is about one-third of all food produced, is either lost or wasted globally (FAO 2011). These postharvest losses present an enormous challenge in order to have enough food to feed an estimated global population of 9.1 billion people by 2050 (FAO 2009). Future demographics will push higher demand for food supplies, which in turn intensifies the need for increasing agricultural yields that could eventually become unsustainable. Reduction of the postharvest losses is being considered as one of the sustainable ways to ensure world food security. Significant efforts have been made to identify the core causes of the postharvest losses, and there are many available techniques and technologies to control these losses. Appropriate low cost innovations and indigenous practices and tools for reducing postharvest losses have also been identified in various countries around the world. But effective awareness of these techniques and proper training in improved postharvest handling are still critically lacking, especially in developing countries and low income nations.

Capacity building, education, and training on factors affecting food loss and wastage and suitable remedies are essential to achieve the goals of reducing food losses and ensuring global food security. This book presents recent advances in postharvest extension/outreach programs, capacity building, and case studies of practical methodologies for postharvest extension professionals and food science teachers, food processing trainers, and outreach specialists who work in the field. Information on the training of postharvest trainers, food loss assessment methods, capacity building at universities and agro-industries, distance education methods, models for cost-effective postharvest/food processing extension work, and success stories and lessons learned from past projects and programs in a wide range of regions, including the United States, Sub-Saharan Africa, South Asia, Central Asia, Latin American, the Caribbean, and Middle East and North African (MENA) nations are described.

The book is divided into four sections. Section I explains postharvest loss assessments methods, Section II is on capacity building, and Sections III and IV focus on training and postharvest extension models. Food loss assessment methodologies are highlighted from several high-profile institutions and it is envisioned that researchers and postharvest extension personnel will benefit from the development and field testing of a hybrid methodology, incorporating the strengths and utilizing the best practices from each of the methodologies in current use.

Several chapters examine decades of the education/extension and capacity building initiatives undertaken in many developing countries, with emphasis on achievements, current and planned activities in postharvest loss assessment, training and extension, postharvest innovation systems, and special events and conferences. Lessons derived from several case studies highlight key considerations with respect to integrating smallholders, gender relations, and nutrition in order to enhance extension/advisory services. Institutions including government and private research centers will benefit from the information from innovative models and methods of monitoring and evaluating projects and programs promoting postharvest loss reduction practices. These will help them to overcome barriers associated with the adoption and calculation of the return on investment of new technologies and practices.

We hope that the information on the postharvest extension/outreach programs, capacity building, and practical methodologies provided in this book will facilitate postharvest extension professionals, trainers, and outreach specialists working in this field to create a cadre of well-trained postharvest specialists. It is expected that these postharvest professionals will further lead their countries towards adoption and implementation of innovative postharvest interventions in order to reduce food losses, increase food availability, and ensure food security for those crops deemed most important to their farmers and citizens.

We would like to express our sincere gratitude to all the authors for their efforts. We appreciate their time dedicated to contributing the chapters presented in the book. We thank the Board of Directors of The Postharvest Education Foundation for their written and editorial contributions, as well as technical support for completing this book. We also thank the staff of CRC Press who has been very helpful throughout the book project.

REFERENCES

FAO. (2009). *How to Feed the World in 2050*. FAO, Rome. http://www.fao.org/fileadmin/templates/wsfs/docs/expert_paper/How_to_Feed_the_World_in_2050.pdf

FAO. (2011). *Global Food Losses and Food Waste: Extent, Causes and Prevention*. FAO, Rome. http://www.fao.org/docrep/014/mb060e/mb060e00.pdf

Editors

Majeed Mohammed is a Professor of Postharvest Physiology at the University of the West Indies (UWI), where he has taught for the past 30 years. After completion of his BSc in agriculture from the UWI in 1977, he obtained his MSc and PhD in postharvest physiology at the University of Guelph (1984) and UWI (1992). He is currently a Board Director with the Postharvest Education Foundation as well as a member of the United Nations/Food and Agriculture Organization Panel of Experts from Latin America and the Caribbean on the Prevention and Reduction of Food Losses and Waste.

Vijay Yadav Tokala is serving on the board of directors for the Postharvest Education Foundation, a non-profit organization committed to reducing global postharvest losses and food waste by organizing and managing postharvest e-learning programs and training activities for the participants from more than 25 different nations. He has worked as horticulture officer for the government of Andhra Pradesh (India) in the roles of field consultant and extension officer, with the main job objective of enhancing quality horticulture production and encouraging farmers toward safe postharvest handling and processing in both rural and urban areas.

He is pursuing a PhD in postharvest horticulture at Curtin University, Western Australia, and studying the effect of novel ethylene antagonists to increase the storage life of fruits in different storage environments. He won the prestigious International Postgraduate Research Scholarship (IPRS)-2016 and Australian Postgraduate Award (APA) to pursue a PhD. His undergraduate studies specialized in horticulture and he was a college topper at Dr. Y. S. R. Horticultural University, India. His postgraduate research focused on fruit and vegetable processing and he was awarded a University Gold Medal for being a topper at S. K. Rajasthan Agricultural University, India.

Contributors

Diane M. Barrett
The Postharvest Education Foundation (PEF),
    USA
Department of Food Science & Technology,
    Emerita
and
University of California
Davis, California, USA

Julia Bello-Bravo
Michigan State University
East Lansing, Michigan, USA

Andrea B. Bohn
University of Illinois at Urbana-Champaign
Urbana, Illinois, USA

Neeru Dubey
Amity International Centre for Postharvest
    Technology and Cold Chain Management
Amity University, Uttar Pradesh
Noida, India

Jorge M. Fonseca
United Nations Food and Agriculture
    Organization (UNFAO)
Rome, Italy

Lola Gaparova
Agricultural University of Tajikistan
Dushanbe, Tajikistan

Deirdre Holcroft
The Postharvest Education Foundation (PEF),
    USA
and
Holcroft Postharvest Consulting
Lectoure, France

Lisa Kitinoja
The Founder, Postharvest Education
    Foundation (PEF)
La Pine, Oregon, USA

B.V.C. Mahajan
Punjab Horticultural Postharvest Technology
    Centre
Punjab Agricultural University
Ludhiana, Punjab, India

Saneya Mohamed Ali El-Neshawy
Professor of Postharvest Diseases
Plant Pathology Research Institute
Agriculture Research Center (ARC)
Giza, Egypt

Radegunda Kessy
International Center for Tropical Agriculture
    (CIAT)
Arusha, Tanzania

Anne Namatsi Lutomia
University of Illinois at Urbana-Champaign
Urbana, Illinois, USA

Roseline Marealle
Independent Consultant
Arusha, Tanzania

Bertha Mjawa
Marketing Infrastructure Value Addition and
    Rural Finance support (MIVARF)
Arusha, Tanzania

Majeed Mohammed
The Postharvest Education Foundation (PEF),
    USA
and
University of the West Indies
Trinidad, West Indies

S. Mohan
Professor of Agricultural Entomology
Tamil Nadu Agricultural University
Coimbatore, India

Ashley Nagele
ADM Institute for the Prevention of
    Postharvest Loss
University of Illinois at Urbana-Champaign
Urbana, Illinois, USA

Ngoni Nenguwo
Independent Consultant
Harare, Zimbabwe

Barry R. Pittendrigh
Michigan State University
East Lansing, Michigan

Sunil Saran
Amity International Centre for Postharvest
    Technology and Cold Chain Management
Amity University, Uttar Pradesh
Noida, India

Sarah Schwartz
ADM Institute for the Prevention of
    Postharvest Loss
University of Illinois at Urbana-Champaign
Urbana, Illinois, USA

Mindy Spencer
ADM Institute for the Prevention of
    Postharvest Loss
University of Illinois at Urbana-Champaign
Urbana, Illinois, USA

Vijay Yadav Tokala
The Postharvest Education Foundation (PEF),
    USA
and
Curtin University, Perth, Western Australia,
    Australia

Karol Alpízar Ugalde
Inter-American Institute for Cooperation on
    Agriculture (IICA)
San Jose, Costa Rica

Charles L. Wilson
Founder, World Food Preservation Center®
    LLC
West Virginia, USA

Alex Winter-Nelson
ADM Institute for the Prevention of
    Postharvest Loss
University of Illinois at Urbana-Champaign
Urbana, Illinois, USA

Devon Zagory
The Postharvest Education Foundation, USA
and
Devon Zagory & Associates
Davis, California, USA

Section I

FLW Assessment Methods

1 Current Status of Food Loss Assessment Measurements and Methodologies

Lisa Kitinoja

CONTENTS

1.1 INTRODUCTION TO FOOD LOSS ASSESSMENT MEASUREMENTS AND METHODOLOGIES

In the past few years, several new food loss assessment methods have been developed and launched by international agencies and global Non-Governmental Organizations (NGOs). These include the World Resources Institute (WRI) Global Food Loss and Waste Protocol; the United Nations (UN) Global Initiative on Food Loss and Waste Reduction (also known as the SAVE FOOD Initiative) field case studies methodology; the International Food Policy Research Institute (IFPRI) Technical Platform on Measurement and Reduction of Food Loss and Waste, which focuses on potential food loss and waste; and an updated Commodity Systems Assessment Methodology (CSAM), led by the

Inter-American Institute for Cooperation on Agriculture. These food loss methodologies are being used to measure postharvest losses from the farm to the consumer and have been developed and promoted for use by field practitioners as an important part of achieving the UN Sustainable Development Goal 12.3, which focuses on reducing food losses/waste by 50% of 2015 values by 2030.

This chapter will introduce these food loss assessment methodologies and provide some background information on their development, differing objectives, applications, and an analysis of their strengths and weaknesses.

1.2   DEFINITIONS OF FOOD LOSS AND FOOD WASTE

Food losses refer to the decrease in edible food mass throughout the different segments of the food supply chains—production, postharvest handling, agro-processing, distribution (wholesale and retail), and consumption. Food losses and their prevention have an impact on the environment, food security and livelihoods for poor people, and economic development. The exact causes of food losses vary throughout the world and are very much dependent on the specific conditions and local situation in a given country, region, or production area (FAO 2015).

An important part of food loss is called food waste, which refers to the removal from the food supply chain of food that is fit for consumption, or which has spoiled or expired, mainly caused by economic behavior, poor stock management or neglect (FAO 2015).

1.3   DESCRIPTION AND ASSESSMENT OF FOUR MAJOR FOOD LOSS ASSESSMENT METHODOLOGIES

There are four commonly used ways to calculate postharvest losses, based on assessments of:

- Physical losses (quantitative losses)
- Loss of calories
- Loss of quality and related changes in market value
- Loss of nutritional value

Food loss assessment provides concrete evidence on the point of loss, amount, causes and sources of food losses, and helps to identify where and when any potential changes in handling practices and postharvest technologies could reduce food losses. Postharvest food losses and waste can be due to:

- Physical damage (rough handling, poor quality packages)
- Pests/diseases (insect and rodent attack, decay/rot)
- Poor temperature and relative humidity management
- Delays (lack of access to transport and markets, poor transportation)
- Deterioration (natural ripening, poor packaging for processed products)
- Stackburn (grain discolouration due to heat build-up due to extended storage)
- Purchasing and/or cooking too much food

The High Level Panel of Experts (HLPE) (2014) report on food losses and waste (FLW) proposes to disentangle the complexity and diversity of causes by organizing their description into three different levels:

1. "Micro-level" causes of FLW: These are the causes of FLW, at each particular stage of the food chain where FLW occurs, from production to consumption, that result from actions or non-actions of individual actors of the same stage, in response (or not) to external factors.

Micro-level causes of FLW include rough handling, use of cheap but non-protective containers, and the use of traditional storage structures.

2. "Meso-level" causes of FLW: These include secondary causes or structural causes of FLW. A meso-level cause can be found at a stage of the chain other than where FLW happens or can result from how different actors are organized together, of relationships along the food chain, or of the state of infrastructures. Meso-level causes include the use of transport via open loads, and lack of postharvest pest management, each of which can contribute to the existence of micro-level causes.

3. "Macro-level" causes of FLW: This higher level accounts for how food losses and waste can be explained by more systemic issues, such as a malfunctioning food system, or the lack of institutional or policy conditions to facilitate the coordination of actors (including securing contractual relations or enforcing quality standards), and to enable investments and the adoption of good practices. Systemic causes are those that favor the emergence of all the other causes of FLW, including meso and micro-causes. Macro-level causes are a major reason for the global extent of FLW (HLPE 2014).

Four different food loss/waste measurement methodologies are in current use, and each has its strengths and weaknesses.

1.3.1　Global Food Loss and Waste Protocol

1.3.1.1　History and Background Information

The WRI tackled one of the big problems in the field of food loss and waste reduction by developing a protocol for organizing the findings of quantitative FLW measurements and reporting the results. Expert panels of postharvest and food loss/waste researchers worked on the protocol as volunteers and external reviewers for nearly 2 years.

1.3.1.2　Objectives

The main objectives are (1) to facilitate the quantification of FLW (what to measure and how to measure it) and (2) to encourage consistency and transparency of the reported data.

1.3.1.3　Strengths

The FLW Protocol is provided as an open source document, free to the public. Its use is being encouraged at the global level by WRI and Champions 12.3, and it has been adopted by several major organizations.

1.3.1.4　Weaknesses

The FLW protocol does not provide a specific quantitative measurement method for researchers to utilize, but only provides guidance on the selection among many different types of measurement methods.

1.3.2　Field Case Studies Methodology

1.3.2.1　History and Background Information

The Food and Agriculture Organization (FAO) via the SAVE FOOD Initiative have developed a "food loss analysis" or field case studies methodology that is being field-trialed and promoted for measuring food losses at critical loss points in food supply chains. An earlier version of the methodology was known as the Informal Fish Loss Assessment Method (IFLAM). This method is a form of rapid rural appraisal (RRA) that was developed and tested by researchers at the UN FAO for measuring food losses in small-scale fisheries. Recently it was combined with Load Tracking (LT) and the Questionnaire Loss Assessment Method (QLAM) and field tested as a package of methodologies. IFLAM is used to generate qualitative and indicative quantitative postharvest fish loss data that can be used to inform decision-making or to plan the use of LT and the QLAM. IFLAM can be used for exploring a food supply chain (FSC) about which little is known. LT is used to quantify and describe losses at stages along the distribution chain or losses related to specific activities, such as harvesting, transport, processing, and marketing. Key data related to the cause and effects of losses from an IFLAM and/or LT study are validated using the QLAM before any suitable intervention is introduced. A combination of the IFLAM, LT, and QLAM can then be used to monitor and evaluate the effects of an intervention intended to reduce food losses (Diei-Ouadi and Mgawe 2011).

1.3.2.2   Objectives

The field case studies methodology has three objectives: (1) to describe the food supply chain, (2) to identify key losses in terms of quantity and quality, and (3) to identify and characterize cost-effective solutions for reducing food losses.

1.3.2.3   Strengths

This methodology is supported by a written manual with clear instructions on data collection, analyses, and reporting (FAO 2016). Both quantitative and qualitative data are collected using mixed methods. Reports based on this methodology are standardized and easy to understand.

1.3.2.4   Weaknesses

The data collection processes (known as screening, survey, and sampling) are ad hoc, wherein the field team can create their own data collection processes and use any practices with which they are familiar. The focus on critical loss points allows the team to focus their efforts on the priority problems in the FSC, but usually does not allow a computation of total losses.

1.3.3   TECHNICAL PLATFORM ON MEASUREMENT AND REDUCTION OF FOOD LOSS AND WASTE

1.3.3.1   History and Background Information

FAO and IFPRI have combined forces to develop a comprehensive food loss protocol that includes the measurement of field-level losses (lost yield or potential food loss) as well as postharvest losses.

1.3.3.2   Objectives

The objective of this methodology is to account for Potential Food Loss and Waste (PFLW) along the value chain.

1.3.3.3   Strengths

One of the components of this methodology is to include statistical analyses of large quantities of field data, and to provide a mobile application for data analysis and reporting.

1.3.3.4   Weaknesses

This methodology is still under development and is currently being field-trialed. The published reports that have utilized the method to date have each used different approaches and definitions of the stages of the FSC, so it is difficult to draw any conclusions.

1.3.4 COMMODITY SYSTEMS ASSESSMENT METHODOLOGY (CSAM)

1.3.4.1 History and Background Information

This postharvest food loss assessment method sets the stage for productive postharvest extension work by assessing the technical, socioeconomic, cultural, and institutional factors related to handling a given commodity in a specific locale. The end products of CSAM encompass both traditional loss assessment and cost-benefit analysis.

The commodity system is made up of 26 components that together account for all the steps associated with the production, postharvest handling, and marketing of any given commodity. The method was developed over the course of many years and was tested extensively in the Caribbean before being introduced worldwide to field personnel via a comprehensive training manual (LaGra 1990). CSAM is being used by postharvest researchers and field level practitioners in more than 30 countries for assessing postharvest losses in fruits, vegetables, root and tuber crops, and staple crops. It is one of the postharvest loss assessment methods taught by The Postharvest Education Foundation (PEF) via its Global Postharvest E-learning Program. The Inter-American Institute for Cooperation on Agriculture (IICA) and PEF have expanded the original CSAM manual to include assessment details for cereals and pulses (LaGra et al. 2016).

1.3.4.2 Objectives

The CSAM process has four objectives: (1) to characterize the commodity system from farm to market, (2) to identify causes and sources of food loss/waste, (3) to measure losses in terms of quantity, quality, and market value, and (4) to identify research needs, training needs, and advocacy issues that, if addressed, will enable reduction of food losses.

1.3.4.3 Strengths

This methodology is supported by a written manual, available for free in digital format. It has been field-tested in many countries over the course of 25 years. CSAM can assist a postharvest loss assessment team to determine the sources of food losses (when, where, and who within the marketing chain is responsible); the causes of those losses (what handling or marketing practices are responsible); and the economic value of the losses compared to the costs of current and proposed postharvest practices.

CSAM provides summary information for the crop on:

- Priority losses
- Causes and sources of losses
- Economic value of losses
- Research needs
- Extension/training needs
- Advocacy needs

1.3.4.4 Weaknesses

CSAM is a case study, rapid assessment methodology, and as such it is difficult to replicate the results over time. Recent studies in Rwanda will provide the first view of repeated CSAM studies (in 2009, 2016, and 2017) for the case of open-field tomatoes.

Table 1.1 is a summary of the four major methodologies, summarizing their objectives, relative strengths and weaknesses, and providing an overview for practitioners who seek to identify and use a method that will best achieve their own objectives.

TABLE 1.1

Comparison of Four Major Food Loss Assessment Methodologies

Methodology	Objectives	Strengths	Weaknesses
WRI-FLW protocol	• Facilitate the quantification of FLW (what to measure and how to measure it) • Encourage consistency and transparency of the reported data	• Standardizes reporting of FLW • Includes farm, postharvest, storage, markets, and consumption • Tracks destinations for lost food • Provides a data-reporting spreadsheet	• Based on haphazard and ad hoc methods for measuring losses and waste, since the emphasis is on reporting FLW • Requires regular inputs from widely varying sources
FAO Field case studies	• Describe the food supply chain • Identify key losses in terms of quantity and quality • Identify and characterize cost-effective solutions for reducing food losses	• Focuses on food losses at critical loss points • Uses a combination of quantitative and qualitative data • Based on a published FAO methodology • Includes blank tables for reporting (2015)	• Still in field trial stage • Case study "snap-shot" approach • Difficult to replicate • Does not include measurements of consumer food waste • Provides a structure but requires the development of your own surveys and measurement protocols
IFPRI PFLW	• Account for PFLW along the value chain	• Uses a combination of quantitative and qualitative data • Includes "potential" FLW as well as measured losses (lost yield) • Statistical analyses of data	• Still in field trial stage
CSAM	• Characterize the commodity system from farm to market • Identify causes and sources of food loss/ waste • Measure losses in terms of quantity, quality, and market value • Identify research needs, training needs, and advocacy issues that, if addressed, will enable reduction of food losses	• Has been in use since the late 1980s • Focuses on postharvest losses • Rapid assessment • Uses a combination of quantitative and qualitative data • Based on a published training manual (2016) • Includes data collection worksheets, sample questionnaires, and reporting outlines	• Case study "snapshot" approach • Difficult to replicate • Does not include measurements of consumer food waste

1.4   STATUS OF FOOD LOSS ASSESSMENT MEASUREMENTS

These four methodologies have been used in various countries to measure food losses for a variety of crops and food products. The majority of the published studies have been focused on horticultural crops and conducted by researchers associated with World Food Logistics Organization (WFLO) or the FAO (Table 1.2).

TABLE 1.2

Examples of the Recent Applications of Methodologies in the Field

Methodology	Country	Crops and/or Food Products	Date of Assessments	References
WRI FLW protocol	Global	All foods	2014	WRI (2014)
WRI FLW protocol	Pakistan	Dairy	2017	WRI (2017)
Field case studies	Kenya	Banana, maize, fish, milk	2014	SAVE FOOD (2014)
Field case studies	Trinidad and Tobago	Mangoes, pumpkin, cassava, hot pepper, tomato	2013–2015	Mohammed (2013) Mohammed and Craig (2014) Craig et al. (2015)
Field case studies	India	Rice, mangoes, chickpeas	2016	SAVE FOOD (2017)
IFPRI PFLW	Ghana	Beans, maize, potatoes, wheat	2016	IFPRI (2016)
PFLW	Peru, Honduras, Guatemala	Potatoes, Maize	2016	Delgado et al. (2017)
CSAM	Rwanda	Tomato, amaranth, banana,	2009	WFLO (2010)
CSAM	India	10 horticultural crops	2009	WFLO (2010)
CSAM	Benin	Tomato, amaranth, orange, pineapple	2009	WFLO (2010)
CSAM	Ghana	Tomato, okra, cabbage	2009	WFLO (2010)
CSAM	Egypt	Tomato	2015	Kitinoja et al. (2016) for Committee for Economic and Commercial Cooperation of the Organization of the Islamic Cooperation (COMCEC)
CSAM	Nigeria	Cassava, sweet potato	2015	Kitinoja et al. (2016) for COMCEC
CSAM	Uganda	Maize, banana	2015	Kitinoja et al. (2016) for COMCEC
CSAM	Rwanda	Tomato, cooking banana, green chilli, sweet potato	2016–2017	Agribusiness Associates Inc. for Horticulture[#] Innovation Lab
CSAM	USA	Tomato, romaine lettuce, peach, potato	2017	Global Cold Chain Alliance (GCCA) for World Wildlife Fund (WWF)[#]

[#] In progress, not yet published.

1.5 CONCLUSIONS AND RECOMMENDATIONS

As described in this chapter, a variety of food loss assessment methodologies are being used to assess and measure FLW, but there are many gaps in the coverage of countries, crops, and food products, and obvious weaknesses associated with each methodology. FLW reports provide data in all kinds of formats and units of measure. Older methodologies and ad hoc methods for measuring food losses often report only on quantitative losses, while these new methodologies also measure qualitative and economic losses. Researchers and practitioners would benefit from the development and field-testing of a hybrid methodology, incorporating the strengths and utilizing the best practices from each of the methodologies in current use. Starting with high-quality FLW data is the best way to make sure that food loss reduction efforts will target the most important postharvest problems.

REFERENCES

Craig, K., Mohammed, M., Mpagalile, J., and Lopez, V. "UN FAO Study on Postharvest Losses in Trinidad and Tobago, Guyana and St. Lucia: Marketing and Economics." *Poster Presented at First International Congress on Postharvest Loss Prevention: Measurement Approaches and Intervention Strategies for Smallholders*, Rome, Italy, October 4–7, 2015.

Delgado, L., Schuster, M., & Torero, M., Reality of Food Losses: A New Measurement Methodology. Consultative Group on International Agricultural Research (CGIAR), 2017.

Diei-Ouadi, Y. and Mgawe, Y.I., "Postharvest Fish Loss Assessment in Small-Scale Fisheries: A Guide for the Extension Officer," *FAO Fisheries and Aquaculture Technical Paper*, no. 559 (2011): 93, http://www.fao.org/3/a-i2241e.pdf.

FAO, *Global Initiative on Food Loss and Waste Reduction: Definitional Framework of Food Loss*, 2015, http://www.thinkeatsave.org/docs/FLW-Definition-and-Scope-version-2015.pdf.

FAO, *Food Loss Analysis: Causes and Solutions; Case Studies in the Small-Scale Agriculture and Fisheries Subsectors—Methodology*, 2016, http://www.fao.org/3/a-az568e.pdf.

HLPE, *Food Losses and Waste in the Context of Sustainable Food Systems*, 2014, FAO, Rome, Italy.

IFPRI, *The African Growth and Development Policy Modeling Consortium*, 2016, http://www.agrodep.org/event/2016-training-course-losses-along-food-value-chains.

Kitinoja, L., Hell, K., Chahine, H., and Brondy, A., *Reducing On-Farm Losses in the OIC Member Countries*, WFLO/PEF Commissioned report for COMCEC, 2016, http://www.kalkinma.gov.tr/Lists/Yaynlar/Attachments/697/Reducing%20On-Farm%20Food%20Losses%20in%20the%20OIC%20Member%20Countires.pdf.

LaGra, J., *A Commodity System Assessment Methodology for Problem and Project Identification* (Moscow, ID: Postharvest Institute for Perishables, 1990).

LaGra, J., Kitinoja, L., and Alpizar, K., *Commodity Systems Assessment Methodology for Value Chain Problem and Project Identification: A First Step in Food Loss Reduction* (Costa Rica: IICA, 2016), http://repiica.iica.int/docs/B4232i/B4232i.pdf.

Mohammed, M., *Analysis of the Postharvest Knowledge System: Case Study on Pumpkins in Trinidad and Tobago*, at Postharvest expert meeting CTA headquarters, Wageningen, the Netherlands, August 10–15, 2013: 25.

Mohammed, M. and Craig, K., *UN FAO study on postharvest losses of cassava, tomato and mango in Trinidad and Tobago, Guyana and St. Lucia*, at UN FAO regional forum to raise awareness on the reduction of food losses along the food chain in the CARICOM Sub-Region, United Nation House, Barbados, September 25–26, 2014: 20.

SAVE FOOD, Food Loss Assessments: Causes and solutions; Case Studies in Small-Scale Agriculture and Fisheries Subsectors—Kenya (Bananas, Maize, Milk, Fish), 2014, http://www.fao.org/fileadmin/user_upload/save-food/PDF/Kenya_Food_Loss_Studies.pdf.

SAVE FOOD, *SAVE FOOD Study in India*: Four Case Studies by Sathguru Consultancy, 2017, https://www.save-food.org/cgi-bin/md_interpack/lib/pub/tt.cgi/India_Project.html?oid=62956&lang=2&ticket=g_u_e_s_t.

Schuster, M. and Torero, M., "Toward a Sustainable Food System: Reducing Food Loss and Waste," chap. 3 in *2016 Global Food Policy Report* (Washington, DC: International Food Policy Research Institute (IFPRI), 2016), 22–31. http://ebrary.ifpri.org/cdm/ref/collection/p15738coll2/id/130211.

WFLO, *Identification of Appropriate Postharvest Technologies for Improving Market Access and Incomes for Small Horticultural Farmers in Sub-Saharan Africa and South Asia* (WFLO Grant Final Report to the Bill & Melinda Gates Foundation, March 2010: 318 pp.)

WRI, *Food loss and waste accounting and reporting standard*, FLW Protocol Case Studies, 2014, http://www.wri.org/sites/default/files/REP_FLW_Standard.pdf.

WRI, *Nestlé Dairy Factories in Pakistan: Losses Across the Value Chain; A Case Study*, FLW protocol case studies, 2017, http://flwprotocol.org/case-studies/nestle-dairy-factories-pakistan-losses-across-value-chain/.

2 Capacity Building in Postharvest Loss Assessment, Postharvest Training, and Innovations for Reducing Losses
Challenges and Opportunities in the Caribbean

Majeed Mohammed and Lisa Kitinoja

CONTENTS

2.1 OVERVIEW OF THE EXTENT OF POSTHARVEST LOSSES IN THE CARIBBEAN

A high incidence of postharvest losses (PHL) exacerbates the problems of low agricultural productivity and food security in countries of the Caribbean Community (CARICOM) (FAO 2012). PHL cause severe reductions in the quality and quantity of foods, thereby affecting incomes and impacting food security in the region. FAO (2012) indicated that PHL are highest in developing countries. Fonseca and Vergara (2013) reported that in the Latin America and Caribbean (LAC) region, 50% of the fruits and vegetables and 37% of roots and tubers are lost before they reach consumers, and articulated further that improving logistics systems and management would be an efficient approach to reduce losses across the supply chain. They found that failure in logistics operations including product handling, pre-cooling, packaging, storage, transportation, and inappropriate infrastructure are among the most common reasons for the high quantities of food losses. These estimates do not include losses in quality or nutritional value, or the health burden associated with consuming contaminated food products.

Several factors contribute to PHL along the supply chain, such as preharvest factors, environmental hazards (inadequate temperature and relative humidity control), pests and diseases, and senescence. Reducing the incidence of PHL along the food chain in the CARICOM subregion will contribute to: improving food availability to address food insecurity, enhancing food quality (better packaging, handling, and storage), increasing economic access to food through job creation and income generation, and development of efficient logistics systems to improve market access by delivering the right product at the right time (FAO 2012).

Efforts to combat PHL in the past have not been very successful, partly due to the fact that countries lack the required and up-to-date information about the scale of the problem. This lack of reliable data on the volume, value, causes, and sources of PHL for crops in the Caribbean region has continuously prevented governments, the private sector, and other key stakeholders from implementing workable solutions to address the problem. While there is an increase in the acknowledgement amongst governments in the CARICOM region and the international community that PHL reduction is one of the key elements in ensuring food security, the use of inappropriate and outdated approaches remains a hindering factor for the current interventions.

2.2 SIGNIFICANCE OF POSTHARVEST EXTENSION

According to Kitinoja et al. (2011), postharvest outreach or extension involves making the link between researchers, producers, and marketers. As such, a priority area to reduce losses of fruits, vegetables, and root crops after harvest lies in the assessment of losses, understanding of the causes and sources of losses, and adaptation, development, and communication of technologies needed for the improvement of current handling, pre-cooling, packaging, storage, and transportation logistics, particularly at the rural level. The importance of establishing and maintaining a positive link between research and extension efforts in seeking to meet these linked objectives cannot, therefore, be over-emphasized.

Shortage of extension staff in the Extension Division throughout the Caribbean is a major constraint to the reduction of PHL and waste. In reality, existing agricultural extension services are entirely directed to production-related problems. Moreover, extension assistants and officers have limited knowledge of postharvest technology. It is essential, therefore, that training in postharvest technology be given to those extension workers who are in direct contact with operations in which PHL are significant. This increased volume of extension work, however, requires extra numbers in the extension staff. According to Kitinoja et al. (2011), an integration of postharvest science, education, and extension services into agricultural curriculum is required on a consistent basis to build a cadre of trained postharvest subject matter specialists in the Ministries of Agriculture in each country within the Caribbean. This initiative could be justified based on the urgent need for capacity building related to postharvest policy formulation at the government level, research, design of appropriate technology, and extension services.

Present-day agricultural extension, however, deals almost exclusively with the primary producer. Accordingly, postharvest extension requires a rather different focus and level of sophistication. This is due to the complexity of the postharvest extension services in view of the diverse range of stakeholders with decision-making power that impact on the magnitude of PHL. These include farmers, processors, importers, exporters and suppliers, machine and transport service providers, packinghouse equipment installers and maintenance personnel, postharvest treatment agents, storage operators, packaging experts, and marketing channel actors, as well as institutions and researchers. The Kitinoja et al. (2011) recommendation for developing countries to develop "Postharvest Training and Services Centers" for measuring PHL, determining their causes, and identifying solutions incorporating innovations under local conditions is timely and urgent. This would ensure that a practical, applied "extension approach," rather than an introspective research approach, is adopted.

Extension work is a two-way process. Farmers' real problems must be discovered and conveyed to the research officer, followed by identification and feedback of the correct answers after proper testing. For the subsistence farmer, his postharvest problems may not be completely technical, but perhaps may also have a political, social, or economic component. It is critical, therefore, for the extension worker to direct each type of problem to the appropriate specialist to blend the various answers into a whole before he or she can provide the farmer with a remedy that can be applied. The researcher must try to ensure that he or she is working on the farmers' major scientific and technical problems, while at the same time considering the political, social, and economic implications when deciding any priorities. It is essential that those involved in the postharvest sector take a very close look at the extension problems involved, considering all possible factors. Failure to understand these complex problems could mean that most of their efforts to manage small farmer losses may well be misdirected and, therefore, a waste of time and money.

2.3 NEED FOR POSTHARVEST EXTENSION AND LINKAGES IN FOOD LOSS REDUCTION INITIATIVES

McNamara and Tata (2015) argued that the extension activities involving PHL prevention face multidimensional challenges, which indicates that a multipronged extension approach is required to reach the various decision-makers and actors involved. In earlier studies, Tyler and Boxall (1984) also emphasized that, in view of the scarce resources that governments and donors encounter, there would be challenges in targeting research, education, demonstration, and extension funds for a PHL prevention program.

Bauer et al. (2009) stated that extension should be the central and integrating element throughout the supply chain in a program to reduce PHL. McNamara and Tata (2015) indicated that extension explains the needs of the farmers to actors and agencies, and in turn informs the farmer of available opportunities. In most Caribbean countries there is a shift from the traditional markets, where produce displays are under ambient conditions, sometimes without shade for extended periods, to fresh produce displays under refrigerated conditions at supermarkets, both within and without chain stores. In both scenarios, the extension service is absent. Produce managers or vendors usually rely on their own knowledge to reduce losses, which may not include awareness of temperature effects on the rate of decay losses, or relative humidity effects on the rate of water loss. However, there is urgent need to provide solutions to existing high PHL and to design and implement PHL prevention programs. McNamara and Tata (2015) and Jakku and Thorburn (2010) provided seven principles, abstracted from extension programs linked to PHL prevention that are relevant to Caribbean stakeholders involved in the postharvest handling of fresh commodities:

a. Understanding of the target audience's perspective to assist in delivery of messages, education, and extension efforts that would promote participation in PHL reduction programs to stimulate changes in behavior or investments.
b. Identifying chain actors and system participants at critical points where interest and support can be expressed; for example, encouraging progressive farmers to demonstrate technologies to peers. This has worked successfully in farmer field school projects for onions in Belize and for cabbage in St. Kitts and Nevis (Mohammed and Craig 2014) (Figure 2.1).
c. Providing finance to farmers as an incentive for improved performance that corresponds to PHL reduction goals. In St. Lucia, contracted farmers selling produce to a supermarket conglomerate are provided with credit up to EC$500 or US$189 to

FIGURE 2.1   Farmer field school lecture demonstration in Belize.

purchase plastic crates and other essentials for preharvest and postharvest activities (Mohammed and Craig 2014) (Figures 2.2 and 2.3).

d. Managing for impact by discerning what various partners can contribute to the overall effort of reducing PHL and providing sufficient resources to achieve results.

e. Employing multiple modes of communication via demonstrations (for instance, on harvesting techniques through farmer field school projects), radio and television messages, postharvest commodity management factsheets, training the trainers workshops, videos, print media, and newspaper articles.

f. Focusing on the system approach where farmers become owners of a learning and discovery process with power and agency to investigate and select approaches that will improve their lives.

g. Designing PHL prevention programs to work with other large-scale agricultural initiatives to work under an enabling agricultural sector environment.

FIGURE 2.2   Fresh produce being inspected.

FIGURE 2.3 Produce packed in plastic crates for transport to chain stores.

2.4 CAPACITY BUILDING AND TRAINING OPPORTUNITIES IN THE CARIBBEAN

Capacity building can be defined as (McNamara and Tata 2015):

- The creation of an enabling environment with appropriate policy and legal frameworks
- Institutional development, including community participation (of women in particular)
- Human resources development and strengthening of managerial systems

Capacity building is much more than training, and includes the following:

- Human resource development: the process of equipping individuals with the understanding, skills, and access to information, knowledge, and training that enables them to perform effectively
- Organizational development: the elaboration of management structures, processes, and procedures, not only within organizations but also the management of relationships between the different organizations and sectors (public, private, and community)
- Institutional and legal framework development: making legal and regulatory changes to enable organizations, institutions, and agencies at all levels and in all sectors to enhance their capacities

Capacity building in postharvest research is needed in many developing countries like those in the Caribbean. This need can be addressed through internships, human resource development for extension workers and subsidiary staff members in university postharvest laboratories and research centers, and via access to web-based information and mentoring (Kitinoja et al. 2011).

Capacity building in postharvest technology has been ongoing at the University of the West Indies in teaching and research at the undergraduate and postgraduate levels. In the Faculty of Agriculture, numerous short courses have been delivered on postharvest handling of tropical commodities throughout the Caribbean via the Continuing Education Programme in Agricultural Technology (CEPAT).

From 2013 to 2015, UN FAO implemented a regional project under its Technical Programme titled "Reduction of Postharvest Losses (PHL) along the Food Chain in the CARICOM Sub-region"

(TCP-SLC-3404) in collaboration with the Ministries of Agriculture in the region. This resulted in PHL assessment of three supply chains in three countries; development and dissemination of regionally relevant tools and methodologies for PHL assessments; capacity building for agricultural technical officers on tools for assessing and reducing PHL; and piloting projects for the development and transfer of knowledge of some technical solutions. Following the determination of PHL for cassava, mango, and tomato in Trinidad and Tobago, Guyana, and St. Lucia (Mohammed and Craig 2014), UN FAO proceeded to undertake a series of capacity building training workshops, with each workshop consisting of 5 days and including farmers, research, extension, and marketing officers, packinghouse operators, produce managers, and food service officers from 13 CARICOM countries (Table 2.1).

For each workshop, Mohammed and Craig (2014) elaborated on the development of quality profiles as influenced by preharvest and postharvest factors of biological and nonbiological origin. Types and nature of qualitative and quantitative losses were presented, providing the template to conduct practical sessions where participants were given demonstrations on the use of postharvest equipment to measure specific physical and chemical quality attributes of selected commodities. Demonstrations were given on the use and application of the postharvest equipment, including the refractometer to measure total soluble solids, the Vernier caliper for dimensions, the thermometer/hygrometer for temperature and relative humidity data, the penetrometer for firmness, and the measuring scale for fresh weight and eventual fresh weight losses. Hedonic scales were used to determine marketable quality, produce defects, and color changes in skin and flesh of horticultural produce. Calculation of quality losses at different critical loss points (CLPs), quality management modelling based on simple equations for under quality, and quality value and incentives for quality management were highlighted and formed the basis for group presentations and discussion.

Participants were taken on guided field tours to reconcile their knowledge and apply the information they obtained from both theoretical and practical exercises on display techniques, sorting and grading issues, postharvest dip treatments, packaging and storage techniques, packinghouse design, and process flow. This tour prepared participants to appreciate subsequent presentations examining weaknesses and opportunities to develop improved handling systems for fresh commodities, with emphasis on PHL reduction strategies, including the development of value-added

TABLE 2.1

Series of 5-Day Workshops on Postharvest Technology Funded by UN FAO 2014–2015

S. No.	Country	Crops for PHL Training	# Trained
1	Antigua and Barbuda	Onion, tomato	35
2	The Bahamas	Onion, hot pepper	24
3	Barbados	Cassava, tomato	42
4	Belize	Corn (dry), onion	27
5	Dominica	Dasheen, plantain	71
6	Grenada	Cassava, soursop	44
7	Guyana	Cassava, pineapple	42
8	Jamaica	Irish potato, onion	15
9	St. Kitts and Nevis	Sweet potato, tomato	41
10	St. Lucia	Pineapple, plantain	25
11	St. Vincent/Grenadines	Sweet potato, dasheen	54
12	Suriname	Papaya, yard long beans	40
13	Trinidad and Tobago	Pumpkin, cassava	41
		Total	**501**

food products. Participants also had an opportunity to observe how produce was being transported, as well as the display of fresh produce and the range of value-added locally produced and imported products in the supermarket.

The training and practical skills generated in these workshops provided appropriate and adaptable techniques to build capacity and promote PHL reduction strategies in the Caribbean, with a pool of subject matter specialists capable of contributing to the national and regional effort to improve food security. The workshops focused on making strategies and approaches easily accessible, applicable, and economical in order to enhance their uptake by different stakeholders, including the private sector (Figures 2.4 and 2.5).

FIGURE 2.4   Infield training on onion drying.

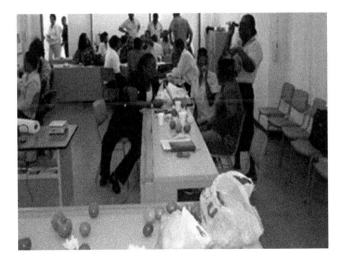

FIGURE 2.5   Quality profile data on tomato in the classroom.

2.5   CASE STUDIES IN BRIDGING GAPS IN POSTHARVEST INNOVATIONS, CAPACITY BUILDING, AND TRAINING

2.5.1   TRINIDAD AND TOBAGO

Throughout the Caribbean, approximately 95% of the small-scale cassava farmers use the garden fork during harvest. Physical losses due to breakages, scaling, cracking, and punctures often promote desiccation, which is accelerated when physically damaged roots are exposed to high temperatures across the handling system, up to retail display. Ultimately, the moisture losses so imposed induce secondary losses due to vascular streaking. Cassava farmer Mr. Ramoutar attempted to reduce such losses by using a mechanical cassava harvester that he fabricated, but losses, arising initially as physical in nature and eventually as physiological, persisted. Based on recommendations from mechanical engineers at the University of the West Indies, the angle of the mechanical cassava harvester rakes was adjusted, significantly reducing physical damages. Mr. Ramoutar was also advised and given several demonstrations to prune the plants one week prior to harvest, and to place harvested roots in plastic crates under shade within one hour of harvest. In the field, washing of roots and subsequent dipping in a fungicide, followed by wrapping in polyethylene bags, were recommended. These recommendations resulted in a drastic reduction in overall PHL to 3%–6%, and a longer retail shelf life of cassava tubers.

In another case study, a farmer who retailed his cassava as a roadside vendor by displaying roots on a table exposed to the sun was advised to invest in an umbrella to reduce the solar heat, which his customers complained often resulted in moisture stress, leading to vascular streaking. The retailer was also given a step-by-step demonstration on how to cover the shaded roots with a moist jute bag, and subsequently advised to weigh and package in low density polyethylene bags only at the point of sale. Encouraged by the increased sales volume arising from a growing number of consumers who recognized higher quality roots with enhanced cooking and baking attributes, the retailer expanded and diversified his operation. Other commodities were displayed for sale, such as sweet potato, dasheen, and plantains. His one day per week retail business changed to three days per week. He even hired a person to assist in the sorting, grading, packaging, and weighing of produce, as well as with monetary transactions. He also interacts with his satisfied customers regularly.

2.5.2   GUYANA

Demonstrations and advice were extended to a fresh produce exporter from Guyana, who focused on the ethnic markets in Toronto, Canada. Guyana has large acreages of mango grown in forested and rural locations throughout the mainland. PHL could be as high as 40%–55% annually (Mohammed and Craig 2014). The exporter was encouraged to negotiate with certain backyard mango tree owners to purchase entire trees, thereby allowing him to harvest mangoes at the mature-green and turning stages of maturity at any time. Mangoes sorted according to these two stages of maturity were placed in plastic crates and transported to the basement of his house, which functioned as his packinghouse and fresh-cut processing facility. Mature-green fruits, as well as fruits with a slight turning but firm in texture, were sliced into 8–10 pieces, washed, rinsed in chlorinated water, then packed in sealed high density polyethylene bags of 5 kg each and frozen at −8°C. Frozen packages were periodically exported in a 40 ft. container to Toronto, Canada. The exporter processed the unsorted, immature mango to make products like chutney, kuchelar, and amchar ,which were sold locally in the municipal markets and supermarkets in Guyana. The ripe fruits were graded and exported, while the overripe fruits were made into jams and jellies for local consumption. By engaging in both fresh ripe fruits and processed value-added products, the exporter significantly reduced PHL to 7%–10%.

2.5.3   SURINAME

Mr. Gopal, a fresh fruit and vegetable exporter who targeted the ethnic markets in the Netherlands, has embarked in a cottage industry in the last 5 years, using the lower floor of his house with an extended shed, to focus on the production of fresh-cut pineapple, watermelon, mango, and papaya for children in the School Feeding Programme in Suriname. With demonstrations and advice on postharvest handling safety and quality protocols pertaining to all aspects of fresh-cut production of fruits from a postharvest consultant, as well as research and extension officers, he secured a soft loan to purchase stainless steel tables, cutting devices, stainless utensils, weighing scales, cold storage rooms, and refrigerated trucks. Mr. Gopal has employed ten women from the village to process his fresh-cut mixed fruit cocktails. Mr. Gopal provides his workers with hair nets, aprons, and training on the significance of sanitation. He has successfully incorporated the cold chain protocols and has a framework to monitor and implement effective quality control.

2.5.4   NEVIS

Mr. Bart is an entrepreneur with a postgraduate degree in Crop Protection, who attended the UN FAO workshop on PHL. His half acre farm is fenced and includes a packinghouse that has an insulated room equipped with a Coolbot™ device attached to the air-conditioned unit. He produces lettuce, cabbage, sweet pepper, pimento, and herbs such as cilantro, parsley, Spanish thyme, and chives. His production is timed to meet supermarket requirements on a consistent weekly basis. Mr. Bart has successfully established all postharvest logistics by harvesting crops during late evening cool temperatures, washing, pre-cooling, packaging in low density polyethylene bags, and overnight storage at 10°C–12°C. The floor is kept wet to keep the relative humidity above 80%–85%. He delivers to two supermarkets within an hour, between 7 and 8 a.m.. Mr. Bart is expanding his operations, and intends to develop value-added products such as bottled green seasoning and pepper sauce. He is in the process of building a more elaborate packinghouse to include more storage rooms, office space, a packingline, and a room to store crates and cardboard cartons.

2.6   RECOMMENDATIONS

Two recommendations are provided for future efforts to reduce gaps in postharvest research, training, extension, and capacity building in the CARICOM region.

1. Include researchers, extension workers, and value chain actors in training of trainer programs in PHL assessment. Having firsthand experience in observing the harvest and postharvest handling, and measuring food losses as the produce moves from the farm to the markets, is a practical way for stakeholders to learn about PHL and share their own knowledge and experiences.
2. Gather current data on PHL for key crops in a wider range of CARICOM countries. This should include information on the amounts and types of losses, the causes and sources of losses, the value of PHL, and the costs and benefits of any potential innovations or interventions intended to reduce PHL. By including researchers, extension workers, and value chain actors in these postharvest studies, local capacity will be increased.

REFERENCES

Bauer, E., Hoffmann, V. and Keller, P. (2009). Principles and guidelines for extension projects. In: *Rural Extension Examples and Background Material*, 3rd ed.; Hoffmann, V., Christinct, A., Lemma., (Eds.)., Weikersheim, Germany: Margraf Publishers, 8, 406–409.

FAO. (2012). *Global Food Losses and Food Waste: Extent, Causes and Prevention.* Rome, Italy: UN FAO.

Fonseca, J. and Vergara, N. (2013). Logistics systems needs to scale up reduction of produce losses in the Latin America and Caribbean Region. *Acta Horticulturae*, 1047, 173–181.

Jakku, E. and Thorburn, P. A. (2010). A conceptual framework for guiding and participatory development of agricultural decision support systems. *Agricultural Systems*, 103, 675–682.

Kitinoja, L., Saran, S., Roy, S. K. and Kader, A. (2011). Postharvest technology for developing countries: Challenges and opportunities in research, outreach and advocacy. *Journal of the Science of Food and Agriculture*, 91, 597–603.

McNamara, P. E. and Tata, J. S. (2015). Principles of designing and implementing agricultural extension programs for reducing postharvest loss. *Agriculture*, 5, 1035–1046.

Mohammed, M. and Craig, K. 2014. Case studies in selected value chains: Postharvest loss management and storage need along the cassava value chain in the Caribbean, Presented at FAO regional conference on cassava in the Caribbean and Latin America, February 10–12, Cave Hill Campus, Barbados: University of the West Indies, 16.

Tyler, P. S. and Boxall, R. A. (1984). Postharvest loss reduction programmes: A decade of activities - what consequences? *Tropical Stored Products Information*, 50, 4–13.

3 Commodity Systems Assessment Methodology
History and Current Improvements

Karol Alpízar Ugalde

CONTENTS

3.1 COMMODITY SYSTEMS ASSESSMENT METHODOLOGY: INTRODUCTION

The Commodity Systems Assessment Methodology (CSAM) helps to identify problems throughout agricultural value chains that cause postharvest losses (food losses) and, at the same time, identify solutions and prepare proposals for improving their efficiency (La Gra et al. 2016).

Using the methodology and the instruments presented in the CSAM manual, it is possible to collect accurate information regarding the agricultural value chain under study. Working together, as an interdisciplinary team, professionals are able to systematically organize their knowledge into a comprehensive overview of a particular agricultural value chain. This will produce the necessary information for the quality problem and project identification.

CSAM makes the basic assumption that human resources are available in all countries, which, on availability of good baseline information on a particular agricultural value chain, could identify projects and establish correct priorities. CSAM provides a complete and accurate information base, which permits users to make the right decisions to overcome the identified problems and implement cost-effective solutions.

CSAM is useful to planning organizations, ministries of agriculture, marketing boards, corporations, research institutes, and other national institutions seeking systematic improvement within existing agricultural value chains. At the regional or national level, the methodology is valuable in the identification and formulation of agricultural development projects. CSAM helps to execute rapid assessment exercises by interdisciplinary teams of national specialists (La Gra et al. 2016).

The application of the methodology requires an interdisciplinary or team approach. It is not possible that one person has all the knowledge to identify correctly the problems related to preproduction, production, harvest, postharvest, and marketing, which are components of any agricultural value chain.

A systematic and interdisciplinary application of this methodology permits a rapid appraisal of an agricultural value chain. It facilitates the identification of problems, their causes, and project ideas. CSAM also helps to provide solutions in development of a strategy and time frame.

An important characteristic of CSAM is that it permits an analysis of the whole agricultural value chain, as well as certain components of interest, thereby facilitating the identification and prioritization of problems throughout the chain. This permits the development of more accurate solutions to priority problems. The methodology brings many concepts, instruments, and techniques together and presents them as an integrated whole (La Gra et al. 2016).

3.2   ORIGIN OF THE METHODOLOGY

The methodology is the result of the work of a number of specialists, and was developed over many years. The original idea for the methodology comes from a study executed in Haiti describing the production and marketing system for beans (*Phaseolus vulgaris*), using an anthropological case study approach (Murray and Alvarez 1973). This case study focused on the varied participants in a particular agricultural value chain and their decision-making processes. It helped as a model for a series of marketing studies carried out in Haiti and in the Dominican Republic by the Inter-American Institute for Cooperation on Agriculture (IICA).

In 1975, IICA's specialists developed a technological approach for describing a food system, incorporating the industrial flow diagram concept with a step-by-step case study method (Amezquita and La Gra 1979). Case studies using this focus were carried out in the Dominican Republic on white potatoes (*Solanum tuberosum*) and tomatoes (*Solanum lycopersicum*) (SEA and IICA 1976, 1977).

After an evaluation of the alternative approaches used by anthropologists, food technologists, and agricultural economists, it became apparent that none of the foci provided a comprehensive picture of an agricultural value chain. However, the integration of the three approaches yielded a complete overview, facilitating problem and project identification.

In 1983, the Postharvest Institute for Perishables (PIP) at the University of Idaho solicited the assistance of IICA (and Dr. Jerry La Gra as IICA's representative) to develop a methodology for quantifying postharvest losses (PHL). The first activity was the application of a modified version of an IICA case study methodology (Amezquita and La Gra 1979) on salad tomatoes and on Chinese cabbage (*Brassica rapa*) in Taiwan (La Gra et al. 1983), under the sponsorship of the Asian Vegetable Research and Development Center (AVRDC).

From the experience in Taiwan, IICA and PIP concluded that loss assessments should begin with a comprehensive overview of the agricultural value chain. Furthermore, due to the high cost in time and resources required to accurately quantify losses, such exercises should only be conducted after an initial assessment of an agricultural value chain, or when quantitative data are required to evaluate the economic feasibility of introducing changes. From that point onwards, both institutions decided to concentrate on developing an approach to evaluate the agricultural value chains using existing instruments and methods.

In an effort to develop a comprehensive methodology for analyzing systems from a postharvest point of view, the ASEAN Food Handling Bureau (AFHB), PIP, and IICA formed an interdisciplinary team in 1986 to visit ASEAN countries and identify problems and needs of public and private institutions dealing with postharvest problems. As a result of numerous sessions with professionals in five countries, the first version of the CSAM manual was prepared (La Gra et al. 1987).

In 1987, the University of California at Davis and PIP, with support from the United States Agency for International Development (USAID), the Food and Agriculture Organization of the United Nations (FAO) and IICA, organized a training course for 20 technicians from the Eastern Caribbean countries. The training concentrated on methods for reducing PHL in perishables, based on an agricultural value chain approach (PIP/UCDAVIS 1987).

The first edition of the CSAM manual was compiled in 1988 in draft form. It was field tested in Malaysia at the Malaysian Agricultural Research and Development Institute (MARDI), under the sponsorship of MARDI, AFHB, PIP, and IICA. During the two-week workshop, 24 MARDI professionals, covering 12 disciplines, applied the methodology.

During the 1990s, the methodology was utilized by Extension Systems International in a number of USAID- and US Department of Agriculture (USDA)-funded projects in Egypt, India, and Indonesia. In 2005, some sections of the first edition of the CSAM manual were translated into Arabic for training the scientists, extension officers, and farmers in Egypt and Lebanon. Since 2008, CSAM has been utilized by consultants trained by the World Food Logistics Organization (WFLO) to organize workshops and training sessions for scientists, university students, and farmers around the world. Since 2011, The Postharvest Education Foundation (PEF) e-learning programs have trained more than 500 young people in the methodology, principles, and practices. CSAM use has been expanded by PEF from the original focus (fruits and vegetables) to all types of cereals, pulses, roots, tuber crops, and cash crops such as coffee (*Coffea* spp.), and added simple data collection practices for conducting rapid appraisals of food losses for extension project and policy development (La Gra et al. 2016).

3.3 COMPONENTS OF CSAM

The methodology divides the agricultural value chain into 26 components, identifying those that fall into the preharvest versus the postharvest stages, and indicating whether they are related to preproduction, production, postharvest handling, or marketing.

Each one of the components is potentially important because the decisions or actions occurring at a point may affect the production, productivity, quality, or cost of the product at that, or at some later, point in the food system. The components included in the methodology are indicative, and some of them may not be relevant to a certain agricultural value chain; or, according to the existing conditions, it may be necessary to include some other components that were not mentioned in the manual.

Some of the components are of an institutional nature and focus on the role that participants play in the agricultural value chain, while other components are of a functional nature, concentrating on processes or activities that take place at a particular point in an agricultural value chain. There are also certain components that may request statistical or descriptive information, which is considered important for the decision-making processes.

The manual provides a brief description of each component, its importance, and the type of information that should be collected. It also provides users with sample questionnaires for collecting the requested information, as well as worksheets. These worksheets and questionnaires can be modified to fit the specific needs of the product, the geographical location, and the interests of the researcher(s).

Furthermore, CSAM provides tools for identifying problems, conducting analyses, proposing solutions, and establishing priorities; this input is critical to undertaking actions and drafting project proposals and/or extension programs geared toward reducing food losses.

3.4 SECOND EDITION OF THE CSAM MANUAL

The medium-term plan for 2014–2018 of the IICA establishes that one of the contributions of the Institute to its 34 member states is: "to improve institutional capacity to reduce food losses and raw materials throughout the agricultural chains." For this reason, in 2016, IICA led a project along with Dr. Jerry La Gra (now a retired IICA specialist), Dr. Lisa Kitinoja, and PEF to publish the second edition of the CSAM manual.

This second edition of the manual introduces revised and updated content, worksheets, and examples of case studies conducted by PEF and Dr. Kitinoja in Africa and Asia.

The document can be accessed free and is available in English and Spanish at this website: English: http://repiica.iica.int/docs/B4232i/B4232i.pdf and Español: http://repiica.iica.int/docs/B4231e/B4231e.pdf.

3.5 THE USE OF CSAM IN LATIN AMERICA

Since 2015, the IICA has led training in capacity building and the application of CSAM in Latin American countries such as Costa Rica, Peru, Uruguay, and Argentina.

Costa Rica was the first country to hold a training workshop on CSAM, which was facilitated by Dr. Hala Chahine of PEF. This activity was attended by 13 IICA specialists from a number of Latin American and Caribbean countries, representatives of the Costa Rica Institute of Technology, and avocado and tomato producer associations in Costa Rica.

In 2016, IICA implemented CSAM in the agricultural value chains for yellow corn and for lettuce in Peru and Uruguay, respectively. In Peru, the study was carried out in the province of Barranca with the support of the General Directorate of Agricultural Policy (DIGNA) of the Ministry of Agriculture and Irrigation (MINAGRI), as well as two cooperatives from the province, Cooperativa Agraria Norte Chico (COOPANORTE) and Centro Ecuménico de Promoción y Acción Social Norte (CEDEPAS Norte). Yellow corn was selected for the project due to its importance as a source of income for family farmers in Barranca province.

In Uruguay, the methodology was applied in conjunction with the General Directorate of Farms (DIGEGRA) of the Ministry of Livestock, Agriculture, and Fishing (MGAP) and Mercado Modelo (the largest wholesale market of agricultural goods in Uruguay). Given that the greatest amount of losses occurs in leafy vegetables, officials selected lettuce for the project.

In Argentina, IICA, along with the Ministry of Agroindustry (MINAGRO), through the Secretariat of Value-Aggregation of Argentina, and the delegation of the FAO, held a training workshop on capacity building in CSAM in 2017.

The workshop was the result of the joint effort of the three organizations within the framework of the National Program on Food Loss and Waste Reduction carried out by MINAGRO, and the need for a methodology shared by specialists in the country to study agricultural value chains with an emphasis on efficiency and reducing losses. The goal of the program is to steer research and offer improvements in all areas of the agriculture food system.

The workshop was attended by 30 professionals from different fields in public and private entities and civil society that are associated with analyzing different agricultural value chains in more than 10 provinces.

In addition, a workshop was held within the framework of the agreement between IICA and the government of the province of Corrientes entitled "Project on competitiveness and sustainability of the agricultural chain of the Corrientes Green Belt 2016–2018." The aim of the workshop was to conduct work on the topic of food losses along the agricultural chain of the Green Belt of this city, and capacity building in CSAM.

Technicians and specialists of the Ministry of Production of the province of Corrientes, the National Institute of Agricultural Technology (INTA), the Secretariat of Family Agriculture, and the private sector participated in the event and deepened their understanding of the methodology by using the agricultural value chains for chard (*Beta vulgaris* var. *cicla*) and lettuce (*Lactuca sativa*) as examples.

3.6 FUTURE STEPS

In 2017, IICA developed a self-learning online course based on the CSAM methodology. The objectives of the course are the following:

- To identify food losses and their causes and impacts on agricultural value chains
- To outline and describe CSAM
- To identify problems and seek solutions by making use of the instruments proposed in CSAM
- To list the basic elements for drafting a project profile oriented toward decreasing food losses in a specific agricultural value chain

The first edition of the course had 300 participants from different countries of Latin America. The course is free of charge and available in Spanish in the IICA's Virtual Campus. In 2018, a new edition of the course will be carried out.

IICA expects to support more members' states in the CSAM application and building capacity in the methodology in the coming years.

3.7   CONCLUSION

CSAM is a methodology under continuous development. Professionals from different organizations have spent years researching and conducting field trials in many countries. New actors in the agricultural value chains are invited to use the methodology. Likewise, any current users are invited to share their experiences, which, without a doubt, will be a valuable source of information to the application of the methodology in the future, different training activities, and for the third edition of the CSAM manual.

REFERENCES

Amezquita, R. and LaGra, J. (1979). *A Methodological Approach to Identifying and Reducing Postharvest Food Losses*. Inter-American Institute for Cooperation on Agriculture. Santo Domingo, Dominican Republic, IICA, pp. 84.

La Gra, J., Jones, J. and Tsou-Samson, C.S. (1983). *An Integrated Approach to the Study of Postharvest Problems in Tropical Countries: A Case Study in Taiwan*. Moscow, ID, University of Idaho, Postharvest Institute for Perishables, pp. 97.

La Gra, J., Kitinoja, L. and Alpizar, K. (2016). *Commodity Systems Assessment Methodology for Value Chain Problem and Project Identification: A First Step in Food Loss Reduction*. San Jose, CR: IICA, pp. 246.

La Gra, J., Poo-Chow, L. and Haggerty, R. (1987). *A Postharvest Methodology: Commodity Systems Approach for the Identification of Inefficiencies in Food Systems*. Moscow, ID, University of Idaho, Postharvest Institute for Perishables.

Murray, G. and Alvarez, M. (1973). *The Marketing of Beans in Haiti: An Exploratory Study*. Port-au-Prince, Haiti, IICA, pp. 60.

PIP (Postharvest Institute for Perishables, United States)/ UC Davis (University of California at Davis, United States). (1987). *Reduction of Postharvest Losses in Perishable Crops: Final Report (of training course)*. Moscow, ID, University of Idaho, Postharvest Institute for Perishables, pp. 86. No. PIP/UC Davis/ California/August 87/No. 94.

SEA (Secretary of State Agriculture) and IICA (Inter-American Institute for Cooperation on Agriculture, Dominican Republic). (1976). *Estudio sobre pérdidas poscosecha de papa en la República Dominicana*. Santo Domingo, Dominican Republic, pp. 75.

SEA (Secretary of State Agriculture) and IICA (Inter-American Institute for Cooperation on Agriculture, Dominican Republic). (1977). *Estudio sobre pérdidas poscosecha de tomate en la República Dominicana*. Santo Domingo, Dominican Republic, pp. 70.

Section II

Capacity Building

4 Postharvest Education, Training and Capacity Building for Reducing Losses in Plant-Based Food Crops
A Critical Review (2010–2017)

Lisa Kitinoja and Vijay Yadav Tokala

CONTENTS

4.1 INTRODUCTION

Global studies conducted by the United Nations Food and Agriculture Organization (FAO), World Resources Institute (WRI), and the World Bank Group (WBG) have indicated that postharvest losses (PHL) are substantial, and commonly quote assessments that estimate that one-third of agricultural produce is lost and does not reach consumers. The reduction of PHL is being increasingly argued for as a sustainable means to increase food availability and ensure global food security (Kitinoja et al. 2011). Effective extension and training play an important role in building capacity along the value chain by encouraging proper postharvest activities. During the last decade, many research studies have been carried out on PHL and detailed findings presented on their causes and sources. On September 25, 2015, the United Nations General Assembly, in the presence of its 193 member countries, announced the ambitious 2030 agenda for sustainable development, along with a set of bold new Global Goals, which included Sustainable Development Goal (SDG) 12.3, with a target of reducing worldwide food waste at the retail and consumer level, and reducing food losses along production and supply chains (including PHL), by 50%. Amongst several other factors, extensive capacity building around the globe is essential to achieve this goal, with a greater focus on under-developed and developing nations, where PHL have been estimated to reach up to 50% (Gustavsson et al. 2011).

Early literature reviews on PHL, conducted for the Bill and Melinda Gates Foundation by the World Food Logistics Organization (WFLO) in 2009 (Kitinoja et al. 2011) and the Missing Food report (World Bank 2011), found very few postharvest-oriented projects and programs. Since then, many international and national organizations have recognized the importance of PHL and launched numerous new initiatives to reduce them. These initiatives include the formation of different postharvest organizations and implementation of projects for capacity building to reduce losses. Several international events and programs were organized to train and educate farmers and stakeholders involved along the different stages of the supply chain. However, despite initiatives in the development community, there is no systematic loss measurement, critical review and syntheses, or utilization of available abundant research findings on ways to reduce losses.

The present chapter provides an overview of postharvest training programs, organizations, projects, and special events for capacity building. This is based on a series of literature reviews undertaken by WFLO and The Postharvest Education Foundation (PEF) from June to October 2017 for the World Bank.

4.2 POSTHARVEST TRAINING AND EXTENSION OUTREACH

The prime objective of agriculture extension is to transfer information from the available extensive knowledge base and multitude of sources to farmers and other stakeholders involved with production and supply systems (Van den Ban and Hawkins 1996).

Postharvest extension or outreach includes establishing a link between research, growers, and marketers. Outreach activities range from word-of-mouth communication, written materials, field visits, and e-learning programs to using information and communication technology (ICT). In past years, several regional and global outreach projects were initiated to train farmers in ways that reduce PHL and to provide the essential infrastructure required.

Many projects were started with an aim to support small and marginal farmers in the adoption of good agricultural, postharvest management, processing, and marketing practices. Such projects were mostly implemented in economically developing and underdeveloped countries or low and lower-middle income countries, and funded by their national ministries or in collaboration with international organizations. "Morocco-Women Entrepreneurs" (2001–2011), "Bangladesh National Agricultural Technology Project" (2008–2014); Marketing Infrastructure, Value Addition and Rural Infrastructure Project (MIVARF) in Tanzania (2012–ongoing) and Cambodian Horticulture Project for Advancing Income and Nutrition (CHAIN) in Cambodia (2015–2017) are a few examples of such initiatives. These projects were started with the chief objective of enhancing income and living standards of farmers by building capacity through public extension agencies on ways to reduce PHL, improve processing, and develop marketing options. Postharvest Loss Alliance for Nutrition (PLAN), with the support of Global Alliance for Improved Nutrition (GAIN), is currently working with local partners and global sponsors in Nigeria and Indonesia to provide essential knowledge and resources for improved postharvest practices that protect the nutritional value of food crops.

4.2.1 Crop-Based Projects

Based on the staple crop in specific areas, crop-based projects were established to train farmers to enhance production and processing technologies, reduce PHL, and encourage value addition for that particular commodity. These projects aim to achieve sustainable improvements in food security, incomes, and livelihoods of farmers or other stakeholders along the supply chain by upgrading and expanding traditional practices of production, handling, and storage methods. The Cassava Mechanisation and Agro Processing Project (CAMAP) in Uganda and Nigeria (2012–ongoing) is working to establish and develop efficient cassava commodity value chains for a reliable supply of processed products for food and non-food industrial usage. In Malawi, a project run by the Ministry of Agriculture and Food Security trained cassava growers to reduce PHL and enhance their incomes through value addition and agro-processing. It provided a detailed guide to processors, with legal specifications for processed foods to preserve and to meet quality standards for marketing. The "On-Farm Seed Storage Project" implemented in seven Asian and African countries and the Postharvest Handling and Storage (PHHS) Project in Rwanda highlighted the importance of improved grain/seed storage techniques using metallic silos to significantly reduce PHL and encouraged better processing and more value addition in target commodities.

4.2.2 University Involvement

Some universities, independently or in collaboration with other funding agencies, conducted training programs for postharvest trainers and developed training manuals on postharvest handling, quality maintenance, and processing. The University of California, Davis (UC Davis) has published the fifth edition of "Small-Scale Postharvest Handling Practices: A Manual For Horticulture Crops" (Kitinoja and Kader 2015), which covers a wide range of topics with simple techniques for postharvest handling and small-scale processing, and the '.pdf' format of the manual can be downloaded from their website free of charge. Purdue University developed Purdue Improved Cowpea Storage technology known as PICS. These low cost, triple layer hermetic storage bags were tested in a long series of field trials in East and West Africa with good success rates and are now being manufactured and distributed by local companies.

4.2.3 Public Sector and Private Firm Collaborations

Public sector and private firms collaboratively established projects to enhance productivity, reduce PHL, and encourage agro-processing for farmers, while providing essential quality raw material to the private firms. Coca-Cola, through "Project Nurture" in Kenya and Uganda (2010–2015), successfully met its targets to reduce PHL in mango and passionfruit by reaching more than 50,000 farmers. The project purchased mango pulp from local processors, who in turn bought mangoes from smallholder farmers (SHFs). The project involved almost 80% of mango farmers in Kenya and helped them double their incomes while fulfilling 100% of the mango requirements of Coca-Cola by local sources (Deloitte and Touche 2015). Nestlé, in partnership with the International Fertilizer Development Centre (IFDC) through Sorghum and Millet in the Sahel (SMS), is training SHFs in the northwestern region of Nigeria on sustainable farming practices while addressing the recurring challenges of poor farm practices, low crop yields, and PHL seriously affecting food security in Nigeria.

4.2.4 Online Training

E-learning or online training has been proved to be an effective way of reaching enthusiastic trainers, extension agents, and practicing scientists in developing regions. It is one of the cost-effective methods to disseminate information and share experiences among people of different expertise levels. PEF, founded in 2011, has been conducting a variety of postharvest e-learning programs on topics including loss assessments, postharvest handling practices, and small-scale processing for young professionals who work with small-scale farmers in developing countries. The e-learning program includes a set of interactive assignments, self-exams, and free access to postharvest training materials relevant to all those involved in agriculture extension work, farmer training, produce handlers, small-scale food processors, and marketers. The ADM Institute for the Prevention of Postharvest Loss (University of Illinois,) has been offering a free four-week e-course through Coursera since 2015, with more than 7000 participants from around the world having completed the course to date. The Philippine Center for Postharvest Development and Mechanization (PhilMech) maintains a webpage on postharvest technologies for rice, corn, and high-value crops (cashew, cassava, mango, onion, and soybean) and offers the e-course "Saving the Lost Harvest: Introductory Course on Rice Post-Production Technologies." A free e-course on food loss assessment for legumes, grains, root, fruit, and vegetable crops is to be launched soon by the SAVE FOOD initiative through the FAO e-learning portal, focusing on FAO's methodology for field case studies.

4.2.5 Information and Communications Technology

ICT usage has increased considerably in agriculture, enabling extension efforts to reach people through mobile message services. Recent studies in Rwanda, India, and Kenya have demonstrated that using mobile services proves to be most effective in not only gathering data but also reaching a wide population with essential information, which may include alerts on market prices, weather forecasting, or the incidence of pests and diseases (Vangala et al. 2017; Tata and McNamara 2018). SAWBO (Scientific Animations Without Borders) has developed an innovative way of creating short animated videos on good agricultural practices, postharvest handling techniques, and storage practices, and distributes them for free via a website, memory sticks, and video-enabled mobile phones to deliver such content to local users (Bello-Bravo et al. 2013). This initiative plays a critical role in overcoming wide prevalent barriers, such as limited local infrastructure and resources, including the low, or absence of, literacy in national languages and variations in dialects. SAWBO actively invites collaborators to record voice tracks in different languages, dialects, and accents most appropriate to the certain region.

4.2.6 Availability of Postharvest Resources

A good number of training manuals and reviews have been published on PHL assessment methodologies, handling, and storage. The World Food Program (WFP) developed and published a free "Training Manual for Improving Grain Postharvest Handling and Storage" (Hodges and Stathers 2012). The updated version of the Commodity Systems Assessment Methodology (CSAM) manual has details on collecting qualitative data for problem identification and postharvest project development with some sample worksheets for plant-based food crops, summary question lists, and guidance on analysing, interpreting, and presenting findings (LaGra et al. 2016). Kitinoja (2016) provided an overview on the status of rural hunger, innovations, and opportunities for reducing hunger and food loss/waste in a review as a part of Brookings Institution's Ending Rural Hunger project.

Information on postharvest practices and marketing is widely available online on websites maintained by different organizations around the world. The extensive information available ranges from regional market prices of certain crops to the use of internationally relevant best postharvest practices. In India, the e-NAM or the "National Agricultural Market" scheme was launched as an online portal integrating all the Agricultural Produce Marketing Committees (APMCs) across different states in the nation to share information on quality production and promote crop exports. The "e-Farmers Centre" in Kenya is a novel agriculture e-platform with a mission to ensure the availability of information on crop production, postharvest handling, and processing for even the smallest of farmers. The application connects buyers, sellers, and transporters while providing training on PHL reduction practices (http://www.e-farmers.co.ke). Many institutions and organizations are also making available online important postharvest information to everyone free of cost. Some of the examples of these resources can be found through UC Davis (http://postharvest.ucdavis.edu/Library/Publications/), FAO (http://www.fao.org/in-action/inpho/home/en/), and PEF (http://postharvest.org/resources.aspx).

International organizations such as FAO, Global Cold Chain Alliance (GCCA), and the World Food Preservation Center® LLC (WFPC) are networking with international postharvest scholars, universities, donors, financial institutions, and private sector partners to form diversified networks in order to act upon critical food system and food loss issues that impact global food security.

4.2.7 CURRICULUM DEVELOPMENT

The inclusion of postharvest related topics in the curriculum of agriculture universities plays a major role in capacity building, by educating young graduates in developing countries in updated technologies for loss prevention and safe and nutritious food preservation. Universities in India, Ethiopia, Kenya, and other countries are already offering a wide range of postharvest related courses and providing scope not only for effective extension but also for further postharvest research and development for innovations relevant to their countries. The WFPC, through its network of '28 sister universities and three agricultural research institutes', has structured different topics to develop postharvest programs in these universities and is also exploring scholarship opportunities for young graduates and postgraduates. The WFPC is also extending support to include relevant topics on food loss, waste, safety, and nutrition to secondary schools in African, Asian, and Latin American nations.

Current education and training on the measurements of PHL are plagued by the use of differing definitions, scopes, and ad hoc data collection methodologies. Standardization of methods and field testing to provide strong evidence for the use of more reliable measurements of quantitative, qualitative, and economic losses are areas that require much attention.

4.3 POSTHARVEST INNOVATION SYSTEMS FOR EXTENSION AND CAPACITY BUILDING

There have been many improvements taking place recently in developing nations related to postharvest innovation systems for education and capacity building. Several major organizations have been involved in postharvest outreach through research and extension, developing new technologies, and providing resources and training to practitioners.

Deloitte and Touche (2015) conducted an extensive study on PHL and value chains for the Rockefeller Foundation, in addition to case studies in Kenya and Mozambique, and have concluded that in order to reduce PHL, the following four components are essential.

1. *Technologies*: Cost-effective technologies for improving postharvest handling, storage, and processing of crops can reduce PHL and provide scope for safe storage until proper market linkages are available. The technology developed should be able to perform well under different conditions, be economically feasible, and adoptable without affecting local social factors.
2. *Market demand and linkages*: Linkages to demand, whether through traditional market relationships or newer procurement and sourcing channels, form the basis of any intervention and provide the foundation for other key elements.
3. *SHF training and aggregation*: Some of the earlier interventions demonstrated the importance of capacity building and other technology adoption measures for SHFs, in order to address the needs and requirements of producers. Farmer group associations facilitating aggregation of producers play a key role in achieving these. Traders also play a prominent role as links to markets and as information sources.
4. *Access to capital investment*: Financing (including loans and leasing models) is essential to ensure adoption of certain technology to reduce PHL in the value chain. It is also required to fund the scale-up of promising technologies and innovative distribution models.

Since 2010, several projects have been initiated to perform research studies on PHL and have presented detailed findings on the causes and sources of the losses, along with possible solutions.

The FAO SAVE FOOD Initiative developed and field-tested a mixed method of "field case studies methodology" for food loss analysis, focusing on systematic assessments and standardized reporting, solution finding, and feasibility assessments (2013–2016).

4.3.1 Innovations and Adoption

Many innovations to reduce PHL suited to different conditions do exist and will keep appearing. The Global Knowledge Initiative (GKI) and the Rockefeller Foundation conducted an intensive innovation scan for more than 6 months, engaging global experts in the fields of agribusiness, academia, investment, innovation, international development, etc. Through this survey a list of 22 investible innovations that deserve to be further investigated, developed, and practiced (GKI 2017) were identified and characterized (http://globalknowledgeinitiative.org/wp-content/uploads/2016/09/GKI-Innovating-the-Future-of-Food-Systems-Report_October-2017.pdf). This initiative was carried out as the Innovation Partner grantee for The Rockefeller Foundation's YieldWise Initiative of US$130 million to reduce postharvest losses by 50% in key value chains by 2030. Many other studies have also suggested different postharvest technologies, but it must remain clear that the application of a certain postharvest technology cannot be considered as a "silver bullet" that solves all PHL issues. Rather, a collaboration between technology, practical technical knowledge, and management skills and advice/support is required. Issues such as the costs and benefits of a technology, economic feasibility, behavioral change, user willingness to pay, adoption barriers, related impacts on the environment, and economic and social factors are also to be considered for improving the adoption rate of postharvest management systems. Potential technology users must be well aware of the basic "best practices" for reducing losses and of different options for their crops of interest, with the possibility of trying any technology or practice on a small scale to ensure cost-effectiveness and suitability per their location and cultural environment. Van Dijk et al. (2016) explained the adoption of different drying and processing techniques for tomato by using the 4A (Awareness, Advantage, Affordability, and Accessibility) adoption analysis in Rwanda. This analysis method was useful even when there was no scientific evidence of the local costs/benefits of the technologies.

4.3.2 Cold Chain Development

Cold chain management along the value chain plays an important role in reducing losses in perishable crops; however, most of the developing world lacks access to affordable refrigeration systems for pre-cooling, transport, cold storage, or freezing during the postharvest handling and distribution of perishable foods. The chief segments of an integrated cold chain include (1) packing and cooling fresh food products, (2) food processing by freezing, (3) cold storage (short- or long-term warehousing of perishables or frozen foods), (4) distribution (cold transport and temporary warehousing under temperature-controlled conditions), and (5) marketing (refrigerated or freezer storage and displays at wholesale markets, retail markets, and food service operations). Total construction and operational costs for refrigerated systems vary widely based on factors such as capacity, local availability, and resource and labour pricing. The overall PHL are reduced by considerable amounts, but the return on investment (ROI) for cooling infrastructure depends largely on the market value of the perishables being cooled and/or stored, and on the use efficiency of the facility (i.e., whether or not it is being operated at full capacity). Cooling or use of refrigerated conditions in developing countries is a rare phenomenon for local or domestic markets and generally is used only for high-value foods intended for export. But usually many follow-on or alternative uses for domestic marketing arise once the cold chain is developed to handle higher-value perishables. For example, in Egypt and India, some fruits like table grapes with good export potential are harvested and typically kept in cold storage for only a—two- to three-month period; the cold storage remains underutilized during the remainder of the year.

Cold storage space and reefer vans are rented to others to handle and distribute other perishable foods such as fish, meat, vegetables, milk products, etc., for domestic markets, during the exporters' off-season (Kitinoja 2014).

Policy makers in the agriculture, energy, education, and food sectors must work together to promote the use of cold chain technology, and improve logistics, maintenance, and services. The policies should also include the development of suitable infrastructure, education, and management skills, and create sustainable markets for the design, use, and funding of cold chains for reducing perishable food losses. Dearman Engine (a United Kingdom-based company) is working on developing new transport vehicles, powered by renewable liquid air fuel that runs a cryogenic engine, to achieve movement of perishable produce through an integrated chain of refrigerated transport and refrigerated storage. Using the case studies of Tanzania and India, Kitinoja (2014) analysed and highlighted the potential for using liquid air driven cold chains to reduce postharvest food losses, raise farmer income, and protect the environment by improving air quality. Dearman recently announced their plans to work with the Indian National Centre for Cold Chain (NCCD) to introduce "clean cold technology" for establishing mobile and stationary cold chain infrastructure in India, aimed at tackling the food crisis by reducing food loss.

4.3.3 ACCESS TO FUNDS

According to the United Nations, and recently confirmed by International Fund for Agricultural Development (IFAD), the main cause of food losses relates generally to the overall inefficient practices that affect different stages of the supply chain (HPLE 2014). These situations can be related to a lack of essential knowledge about recent relevant technologies amongst extension agents, who form the link between researchers and practitioners. In some developing countries like India and Kenya, good coordination does exist between agricultural universities, research centers, and the extension system, yet a lack of sufficient infrastructure and resources (fuel/electricity) limit an efficient transfer and adoption of the technology. Access to capital, required for access to essential resources, forms a key requirement for wider adoption of certain postharvest technology to reduce losses. Provision of subsidies on essential infrastructural requirements (fuel/electricity), useful technologies, and purchase guarantees can be used to ensure SHFs gain access to technologies. For example, electricity costs were subsidized for commercial users in Trinidad and Tobago at four cents per kilowatt-hour (kWh) in 2016, while subsidies were as high as 22 cents per kWh in Rwanda and up to 38 cents per kWh in Papua New Guinea. In India, the provision of subsidies and technical advice towards infrastructure development of cold storages, packhouses, reefer vans, etc. for farmer groups through a centrally sponsored scheme—the Mission for Integrated Development of Horticulture (MIDH) (http://midh.gov.in/)—is showing positive results in increasing the rate of technology adoption by farmers, serving to reduce PHL and improving marketing of horticultural produce. In Gambia, success of the Gambia Horticultural Enterprise (GHE Ltd.) in implementing projects to promote agro-processing of mango fruits suggested that donor assistance and private sector investment can greatly contribute to the development and rapid expansion of the agriculture sector, increasing foreign exchange earnings, creating employment opportunities, improving food and nutritional security and market outlets, and generating rural incomes and higher living standards for small growers.

4.3.4 POSTHARVEST INNOVATION PLATFORMS

The major role of the postharvest capacity building is to impart information on recent relevant technologies to extension agents, trainers, and practitioners. A postharvest innovation platform involved in capacity building should be equipped with essential infrastructure to provide key technical information to individuals of different backgrounds and interests, including farmers, traders, food processors, researchers, and government officials. The following are some of the essential features of a postharvest innovation platform, as explained by Kitinoja and Barrett (2015):

1. A physical location for extension workers and local postharvest trainers to meet with farmers and other practitioners along the food value chain to provide training to improve local capacity and knowledge on improved production techniques and harvesting, and proper postharvest handling practices such as sorting/grading, packing, cooling, storage, food safety, processing, and marketing practices.
2. A training venue with the permanent setup for demonstrations of improved, cost-effective small-scale postharvest handling practices, facilities, and equipment.
3. A stage for local private firms to allow them to demonstrate benefits and give details of their goods and services related to improved postharvest handling practices, facilities, and equipment.
4. A one-stop retail store for postharvest tools and supplies, packages, plastic crates, and other goods that can be purchased at reasonable cost.
5. Consultancy services allowing people to ask questions and get advice on how to use improved postharvest practices, and learn about marketing options and costs and benefits.
6. Possession of simple postharvest handling and processing machinery and the ability to provide small-scale services, such as packing using improved containers/packages, facilities for cooling and/or storing products for few days before marketing, leasing of small insulated vehicles, value-addition using the solar dryer to make dried fruit and vegetable snacks, etc.

All the postharvest technologies require adequate technical knowledge and appropriate management skills to achieve success. Postharvest innovation platforms such as Sasakawa Africa Association's Postharvest Extension Learning Platform (PHELP) and the United States Agency for International Development (USAID)-sponsored Postharvest Training and Services Center (PTSC) are examples of innovation platforms working effectively to achieve this requirement. PHELP is presently operating in Ethiopia, Mali, and Nigeria, and PTSCs have been launched in Cape Verde (2009), Tanzania (2012), and Rwanda (2017). A Punjab Horticultural Postharvest Technology Center (PHPTC) was designed and set up in India, and was later reorganized and relaunched in 2017. In Nigeria, there is a World Bank-funded Postharvest Center of Excellence at the University of Benue State.

The PHPTC in India was developed to conduct research and train farmers to reduce losses, promote value addition, establish cold chains, and provide consultancy services. The PTSC models in Tanzania showed positive results as an effective postharvest innovation platform integrating the technology and information needed for postharvest investments and development. The complete detail about the PHPTC and the PTSC based at the World Vegetable Center's (WVC) offices in Arusha, Tanzania is discussed in separate chapters in this book. The Market Infrastructure, Value Addition and Rural Infrastructure (MIVARF) value addition centres and Tanzania Horticulture Association (TAHA) farmer service centers are based on the model of the PTSC. Liberia Agribusiness Development Activity (LADA) and AfricaRice have a paddy processing unit and Central Agricultural Research Institute (CARI) has a cassava processing unit under construction near the site where a new PTSC is being developed in Liberia. The project objective of MIVARF is to enhance rural incomes and food security through improved markets by increasing the share of value-added products for small- and medium-scale producers and processors, and by training and matching grants. This project includes 16 district level training centers in Tanzania to train stakeholders for value addition, including the drying and storage of staple crops, the handling and marketing of fresh produce, and the processing of perishable crops. TAHA launched and is managing three farmer training and services centers for small-scale producers of avocados, pineapples, and vegetable crops.

Value chain development projects for high-value crops have been implemented through the development firms Fintrac and Winrock International with funding from international donors like IFAD and USAID. These projects conduct feasibility studies and develop scenarios based on agribusiness principles and business planning. They have identified that levels of PHL are very high, commonly during the glut period when produce supply is far higher than demand. Processing these crops would not only add value to the produce but would also allow them to be preserved for a longer period. Dried tropical fruits are an attractive business opportunity, especially in countries with favorable weather conditions (i.e., dry) during harvest. In Rwanda, solar-dried pineapple is being exported to the European Union and Tanzania. Dried fruit has an attractive market among the international tourists who purchase these for their backpacking and mountain hiking travels. The GHE Ltd., an agribusiness company, implemented two World Bank matching grant projects to facilitate value addition and market access to Gambian mangoes by processing fruit into pulp, jam, juice, and dried flakes, and promoting the export of fresh and processed mango fruits by sea and air to Europe.

To address some of the issues for postharvest handling and to reduce losses, the government of India, through the Ministry of Food Processing Industries, is promoting the innovative concept of "Mega Food Parks" as a regional hub for food processing enterprises, which will include provision of all the essential infrastructure, roads to major markets, and access to cold storage and transport. Mega Food Parks aim to provide a mechanism to link agricultural production to the market by bringing together farmers, processors, and retailers to one location in order to ensure minimal wastage, higher value addition, increased farmer incomes, employment opportunities, and the empowerment of women in the rural sector (http://www.mofpi.nic.in/Schemes/mega-food-parks).

4.4   SPECIAL EVENTS AND POSTHARVEST CONFERENCES

In the past decade, several global and regional postharvest events were organized to bring together the experts and enthusiasts from different corners of the world. They facilitated communication and discussion related to different postharvest issues and forged intellectual relations.

SAVE FOOD is a global joint initiative of the UN FAO, the United Nations Environment Programme (UNEP), Messe Düsseldorf, and INTERPAK, the leading global trade fair for packaging and processes. It regularly conducts workshops, meetings, and events to facilitate collaboration between various actors from public and private sectors, with a motto to reduce food losses, and different themes related to postharvest, private sector support and involvement, and innovations in food technology. Recently, the VI International Postharvest Unlimited Conference in Spain addressed many aspects related to postharvest quality, from breeding to production and postharvest handling, and provided a platform for scientists, professionals, and students to present and exchange ideas and research findings. The Carrier/United Technologies corporation conducted three "World Cold Chain Summits" and invited participants representing public and private sector cold chain operators, suppliers, engineers, and academics to discuss current technologies for reducing food loss and waste. The recent summit in Singapore was attended by 150 delegates from over 35 countries. In the past 2 years, there were many events organised by national and international organizations to address postharvest issues in the African continent. Examples include the Yieldwise Cassava Innovation Challenge (Sub-Saharan Africa), the Third All Africa Horticultural Congress in 2016 (Nigeria), the First All Africa Postharvest Congress and Expo in 2017 (Nairobi, Kenya), and an International Postharvest Conference marking the launch of Ethiopian Society of Postharvest Management (ESPHM) in 2018.

The First International Congress on Postharvest Loss Prevention, Rome was organized by the ADM Institute (ADMI) and Rockefeller Foundation to assess challenges associated with PHL measurements, metrics, and tools, and to discuss interventions to reduce PHL suitable to smallholders in developing countries. The Congress launched an interactive "Postharvest Losses Roadmap" intended to focus the global network on achieving significant reductions by 2050. The second Congress will be held in 2019. WFPC organised the First All Africa Postharvest Congress and Exhibition in Nairobi, Kenya with over 600 in attendance from 40 different countries. The next two conferences are being planned for 2019: the First All Latin America/Caribbean Postharvest

Congress and Exhibition (Bento Gonçalves, Brazil), and the First All Asian Postharvest Congress and Exhibition (Lucknow, India).

Competitions have been conducted to recognize and encourage new postharvest innovations. East African Postharvest Technologies Competition—sponsored by USAID, Inter Region Economic Network, East Africa Trade Investment Hub, and Syngenta—awarded a cash prize to the top three entries; a bench type manual maize sheller was awarded the first prize. Similarly, a Postharvest Innovations Competition was conducted during the First All Africa Postharvest Congress and Exhibition in Kenya to award cash prizes to the top ten innovations, with a low cost "Dry Card" (UC Davis), which measures storage relative humidity, capturing the first prize, followed by cassava peel processing for animal feed (International Institute of Tropical Agriculture [IITA]) and a rapid batch drying technology (Agricultural Cooperative Development International/Volunteers in Cooperative Assistance [ACDI/VOCA] and IITA).

Several other events on food loss assessment, prevention methods, and postharvest technologies and management have been organized, and many more are planned for the near future.

4.5   ANNUAL UPDATES

There is a growing body of global organizations involved in postharvest research, outreach, and policy development work, and our review has identified a large number launched during 2011 to 2017. An internal review of World Bank projects and programs is currently underway, and a Microsoft Excel spreadsheet of Postharvest Events, Organizations, Programs, and Projects has been compiled and uploaded to the PEF website. Readers are encouraged to submit information on their own current projects and upcoming events. (http://postharvest.org/calendar_of_upcoming_programs_and_meetings.aspx) (Table 4.1).

With each month that passes, additional new events are announced, including a International Potato Center (CIP) workshop in Peru (with International Food Policy Research Institute [IFPRI] and Inter-American Institute for Cooperation on Agriculture [IICA], a Postharvest Management Conference and Postharvest Society launch in Ethiopia, a Food Loss and Waste Reduction Program in Mauritius, and

TABLE 4.1

The Growing Body of Global Organizations Involved in Postharvest Research, Outreach, and Policy Development Work (2011 to 2017)

Year	New Postharvest Organizations Formed	New Postharvest Programs Launched	New Postharvest Projects Initiated	Postharvest Loss Reduction Themed Events Held
Pre-2011 (ongoing through 2011 or 2012)	2	5	12	1
2011	4	2	5	1
2012	1	3	8	1
2013	2		11	
2014	2	4	12	2
2015	2	8	9	8
2016	3	4	9	3
2017	2	5	6	14
2018 (planned)	1	1	1	3
2019 (planned)				2
Subtotals	21	32	73	34

Note: Does not include the World Bank projects, which are still being assessed.

a Clean Cooling Conference at the University of Birmingham for 2018. New projects are also being launched, such as the Postharvest Project in Burkina Faso, which is setting up a PTSC for tomatoes with USAID/Horticulture Innovation Lab funding, and the Postharvest Management in Sub-Saharan Africa (PHM-SSA) Project, now in its second phase, implemented from April 2017 to March 2020. The major emphasis of the second phase of this project is on promoting the adoption of proven technologies and good practices on postharvest management to scale, at both a national and a regional level.

4.6 CONCLUSION

Since 2010, overall awareness regarding PHL has increased considerably and a number of positive developments have been evidenced. Many regional and global outreach projects have been initiated, training different stakeholders of food value chains to reduce PHL, and providing the essential capacity building infrastructure. Several projects were designed with an aim to support small and marginal farmers in economically developing nations towards the utilization of good agricultural practices, postharvest management, processing, and marketing. A good number of postharvest training manuals, research results, and reviews were published and the information on postharvest practices and marketing is now readily available on websites maintained by different organizations around the world. The scope of using ICT to provide effective agriculture extension services is being widely explored, and the use of messaging services in mobile phones and short animated videos has proved to be effective in educating a wide population on different aspects of agriculture and postharvest handling. The inclusion of postharvest related topics in the curriculum of agriculture universities is a good sign for future capacity building. Several projects were initiated to perform research studies on PHL and presented detailed findings on the causes and sources of the losses, along with possible solutions. Yet a lack of sufficient infrastructure and resources (such as fuel, electricity, and credit) are limiting efficient adoption of these technologies. Most of the developing world, in general, still lack access to cold chain infrastructure during the postharvest handling and distribution of perishable foods. Despite significant enthusiasm amongst the development community, a gap in critically reviewing and synthesizing the abundant research on food losses still remains. Current education and training on the measurements of PHL are plagued by the use of differing definitions, scopes, and ad hoc data collection methodologies. Effective education and training to promote capacity building at the national level will require increasing its scope to create a cadre of well-trained professionals. The aim is that these young postharvest specialists will lead their countries toward increased adoption and implementation of existing cost-effective postharvest interventions, while serving as postharvest loss assessment team members, postharvest extension workers, private sector postharvest trainers, consultants on feasibility studies, and postharvest program/project evaluators.

ACKNOWLEDGMENTS

The authors thank Geeta Sethi and Farbod Youssefi of the World Bank Group (Washington, DC) for their feedback on the full set of literature reviews conducted by PEF and WFLO during 2017 for the World Bank, and for reviews of the initial draft of this chapter.

REFERENCES

Bello-Bravo, J., Dannon, E., Agunbiade, T., Tamo, M., and Pittenberg, B.R. (2013). The prospect of animated videos in agriculture and health: A case study in Benin. *International Journal of Education and Development Using Information and Communication Technology*, 9 (3): 4–16.

Deloitte and Touche. (2015). Reducing post-harvest loss through a market-led approach. Retrieved from https://www2.deloitte.com/content/dam/Deloitte/za/Documents/consumer-business/ZA_FL2_ReducingPHLThroughaMarket-LedApproach.pdf

Global Knowledge Initiative (GKI). (2017). Innovating the future of food systems – A global scan for the innovations needed to transform food systems in emerging markets by 2035. Retrieved from http://globalknowledgeinitiative.org/wp-content/uploads/2016/09/GKI-Innovating-the-Future-of-Food-Systems-Report_October-2017.pdf

Gustavsson, J., Cederberg, C., Sonesson, U., van Otterdijk, R., and Meybeck, A. (2011). *Global Food Losses and Food Waste: Extent, Causes and Prevention*. FAO, Rome, Italy.

HLPE. (2014). Food losses and waste in the context of sustainable food systems. A report by the high-level panel of experts on food security and nutrition of the committee on World Food Security, Rome, Italy.

Hodges, R.J. and Stathers, T. (2012). *Training Manual for Improving Grain Postharvest Handling and Storage*. UN World Food Programme (Rome, Italy) and Natural Resources Institute, Gillingham, UK. pp. 246.

Kitinoja, L. (2014). Exploring the potential for cold chain development in emerging and rapidly industrialising economies through liquid air refrigeration technologies. *Liquid Air Energy Network*. 6–17. Retrieved from http://www.liquidair.org.uk/cold-chain-development.pdf

Kitinoja, L. (2016). Innovative approaches to food loss and waste issues, Frontier Issues Brief submitted to the Brookings institution's ending rural hunger project. Retrieved from https://www.endingruralhunger.org/

Kitinoja, L. and Barrett, D.M. (2015). Extension of small-scale postharvest horticulture technologies—A model training and services center. *Agriculture*, 5(3), 441–455.

Kitinoja, L. and Kader, A.A. (2015). Small scale postharvest handling practices: A manual for horticultural crops. Hort Series No. 8E. Davis, CA: UC PTC.

Kitinoja, L., Saran, S., Roy, S.K., and Kader, A.A. (2011). Postharvest technology for developing countries: Challenges and opportunities in research, outreach and advocacy. *Journal of Science of Food and Agriculture*, 91(4), 597–603.

LaGra, J., Kitinoja, L., and Alpizar, K. (2016). *Commodity Systems Assessment Methodology for Value Chain Problem and Project Identification: A First Step in Food Loss Reduction*. San Isidro, Costa Rica: Inter-American Institute for Cooperation on Agriculture.

Tata, J.S., and McNamara, P.E. (2018). Impact of ICT on agricultural extension services delivery: Evidence from the catholic relief services SMART skills and Farmbook project in Kenya. *The Journal of Agricultural Education and Extension*, 24(1), 89–110.

Van den Ban, A.W. and Hawkins, H.S. (1996). *Agricultural Extension*. 2nd ed. Oxford, UK: Blackwell.

Van Dijk, N., Dijkxhoorn, Y., and Van Merrienboer, S. (2016). SMART tomato supply chain analysis for Rwanda: Identifying opportunities for minimizing food losses. SMASH Program, implemented by the Dutch Horticultural Trade Board, BoP Innovation Center, TNO and Wageningen University; financed by the Dutch Ministry of Foreign Affairs.

Vangala, R.N.K., Banerjee, A., and Hiremath, B.N. (2017). Linkage between ICT and agriculture knowledge management process: A case study from non-government organizations (NGOs), India. In: *International conference on social implications of computers in developing countries* (pp. 654–666). Springer, Cham, Switzerland.

World Bank (2011). Missing food: The case of postharvest grain losses in Sub-Saharan Africa. Report No. 60371-AFR. World Bank, Washington, DC.

5 Enhancing Postharvest Capacity in National Extension Systems and Advisory Services

Lisa Kitinoja and Jorge M. Fonseca

CONTENTS

5.1 INTRODUCTION

Rural and agro-industry development in countries partially relies on how research knowledge and innovation are transferred to food producers and processors via outreach and extension services. Indeed, extension services have been considered the main pillar in providing capacity development across the value chain and strengthening the food sector and agricultural business services, including those for postharvest/agroprocessing (Wongtschowski et al. 2013).

Despite the need for generating this "know how" on postharvest systems and agro-industries, most of the extension/advisory service models appear to exclude knowledge transfer and capacity building in areas related to issues affecting the supply chain and business beyond farm gate; rather, they follow a scheme as illustrated in Figure 5.1. Most of the available funding, educational background, and focus of outreach programs are still mostly production-oriented (i.e., seed technology, input technology, and production practices). Moreover, most of these services (public or private) are still centred on the crop/food product itself, while overlooking social and cultural conditions in which producers and agro-businesses interact. The diverse reasons for the lack of postharvest/agro-industry outreach services include, among many: limited budgets dedicated to extension; priorities targeting production; lack of knowledge on how to stimulate private sector investment and services; lack of postharvest and agro-industry technical specialists; and inefficient knowledge transfer from research to an extension.

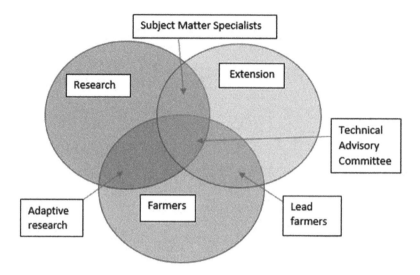

FIGURE 5.1   Traditional model of research extension farmer linkages.

Recent calls for reducing documented high levels of food losses and waste (Gustavsson et al. 2013; Lipinski et al. 2013) have been echoed by many global organizations and agencies. A rapid analysis on the state of innovation in agro-industries conducted in different areas of the developing world (South East Asia, Latin America, and the Caribbean) revealed great disparity in developments within regions (UN FAO, unpublished). Additional extension services or an improved mechanism to transfer knowledge were considered of paramount importance to promote innovation in most successful countries and enable an environment that would allow improved insertion of smallholders, innovation, and overall sector development. As a result, understanding how to establish mechanisms that efficiently enable country advisory services was concluded as an urgent need.

The subsequent step was to undertake a desktop study that would collect information on the historical role of extension/advisory services in achieving the multiple objectives of National Agricultural Development Goals (NADG), including food security, rural livelihoods, natural resources management, social capital, and improved nutrition, as well as how postharvest/agro-industry fits into these key targets. An introduction to the vast and growing literature on global postharvest losses (PHL) and food wastage is used to set the stage. A wide range of examples of national extension/advisory services in various stages of development and integration of postharvest/agro-industry topics and issues were reviewed. This chapter provides a synopsis of the work done, with an emphasis on common factors found in cases where the extension model is showing great performance, given the holistic approach observed.

5.2   OBJECTIVES

1. To describe the current status of postharvest/agro-industry outreach, professional networks, and practical training/educational programs available in the public, private, and global online sectors
2. To provide recommendations on the size, scope, and management of expanded and enhanced extension/advisory services, with guidance on the postharvest/agro-industry related skills, technical expertise, and capacity building that will be needed

5.3   METHODS/APPROACH

The analysis was based on literature reviews, surveys (33 key informants from 20 countries), and case studies undertaken for a 2015 United Nations Food and Agriculture Organization (UN FAO)- Infrastructure and Agro-industries Division report. Commissioned countries were those (1) where extension/advisory services have embraced the challenging issues and problems found in the postharvest/agro-industry sector as a means of promoting food security and rural development (India, Chile, Thailand, the United States, and the European Union countries); (2) that are newly involved in the postharvest sector (Tanzania, Ethiopia, Ghana, Sierra Leone, Nigeria, Cambodia, and Indonesia); and (3) showing recent interest in the concept but not yet making significant investments (several countries in the Middle East Region, Malawi, Nepal, and Rwanda).

5.4   RESULTS

The general perception by those involved in the food sector is that extension services should provide more holistic support. The simple, traditional extension goal (to increase agricultural production) has been widened to include a host of other supporting topics and related skills. Mohamed et al. (1995) conducted a survey of 96 agricultural educators and 128 international graduate students from 33 universities, and noted: "Results indicated that both extension educators and international graduate students of extension education in the United States agreed that the agricultural extension organizations in developing countries should commit their resources to satisfy the educational needs of rural people regarding: (1) application of new inputs, varieties, and improved farm practices, (2) application of new and improved practices related to livestock production, (3) food storage, processing, and preservation, (4) knowledge and skills for family improvement, (5) civic skills, (6) supplementary skills for farm maintenance, and (7) farm business management."

It is understood that extension/advisory services are intended to serve as the link between academic researchers and practitioners/clientele. However, there is often a large educational gap between agricultural/food science researchers (MS and PhD degree holders) and practitioners (who may not be able to read). The extension professional is seen as the potential link between the two groups, by translating scientific information and jargon into practical instruction and demonstrations of "best practices."

5.4.1   Types of Extension/Advisory Services

Swanson and Rajalahti (2010) described four approaches for extension/advisory service work:

1. Top-down technology transfer: often prescriptive or persuasive, an older extension model that relies on a research-based package of crop-specific technical information that is suitable for the user. Primary goal: to increase food production.
2. Problem responsive advisory services: commonly used by private sector extension providers to provide advice based on research, and to supervise contract farmers. Primary goal—to make sure specific inputs are used, and production goals are met.
3. Nonformal education or training: offering agricultural education to organized groups of rural people. According to Swanson and Rajalahti (2010), "This approach continues to be used in most extension systems, but the focus is shifting more toward training farmers how to utilize specific management skills and/or technical knowledge to increase their production efficiency, or to utilize specific management practices, such as integrated pest management (IPM), as taught through Farmer Field Schools (FFS)." Primary goal: to help farmers gain new knowledge and skills.
4. Facilitation extension: market-driven, and works with different groups of farmers (e.g., smallholders, women farmers, landless farmers, youth, etc.) first to identify their specific

needs and interests, and then to identify the best sources of expertise that can help these different groups address specific issues and/or opportunities. Local, more innovative farmers who are already producing and marketing higher value food products, subject matter specialists, researchers, private-sector technicians, and consultants can all be sources of practical information and market linkages. Primary goal: to help groups to develop new products and access new markets.

The analysis of the countries in this study has shown that the emphasis prevails towards approaches one and two, with late improvements in including the third approach, but still in need to improve integration of modality for approach number four.

5.4.2 Inclusiveness in the Extension Service System

If establishing advisory services has already been difficult in many scenarios, it is easy to imagine the little attention at times towards addressing gender equality. Women constitute 43% of the agricultural workforce, producing a large portion of the world's food crops (FAO 2011). Women are active in many aspects of the postharvest chain and food processing industry, as well as playing a major role in food marketing and food preparation in the home. Thus, extension/advisory services must take the specific needs of rural and urban women into consideration when developing outreach programs. Farnworth and Colverson (2015) pointed out that there are often cooperative activities taking place within the family unit (for example, men and women working together on producing and marketing a crop) and recommend "it is useful to think of the extension and advisory services as a facilitation system rather than a service and to reconfigure it accordingly. Existing 'best bet practices' can be captured, integrated, and scaled out to build an empowering extension and advisory facilitation system" in order to reach more women.

Another group that is often marginalized is the small producers. The needs of smallholders have often been neglected due to the difficulties and the high cost of reaching large numbers of people who work in scattered areas on small plots of land. Extension/advisory services in the private sector typically respond to those who can pay for services, and public extension workers can achieve more impact by working with larger scale clientele. Indeed "lead farmers" in the traditional model of extension (Figure 5.1) normally refer to farmers above a certain dimension and capacity. For rural development to succeed, the agro-industry clientele of all scales (small, medium, and large) must be included in value chains.

Given the limitations of extension budgets, one approach is to enhance the capacity of existing extension professionals and practitioners, rather than hiring specialists. Many authors have provided guidance on upgrading the skills of extension workers and better integrating women into the developing food system. In Myanmar, Cho (2002) found that extension agents required in-service training in participatory extension approaches, after years of using the Training & Visit (T&V) system, and generally had poor communications with farmer clientele. Quisumbing and Pandolfelli (2009) and Manfre and Nordehn (2013) reported on how some means of communication (radio, word of mouth) are better than other forms when trying to reach women. Manfre and Sebstad (2011) found similar trends in their research on behavior change in agricultural value chains in Ghana and Kenya.

5.4.3 Transferring Knowledge versus Building Local-Knowledge in Postharvest Technologies

The older extension/advisory services systems had the goal of knowledge transfer, while newer approaches are innovation systems. One example of an innovation platform for postharvest/agro-industry development is the Postharvest Training and Services Center (PTSC), where extension/advisory service workers facilitate learning, trial and adoption of practice changes, and access to information, tools, supplies, and markets for smallholder horticultural clientele.

A central site for conducting postharvest research and offering local extension programs and advisory services, such as a PTSC, is recommended for each and every developing country (Kitinoja and Al-Hassan 2012), as well as for underdeveloped rural communities in developed countries. For example, this model innovation platform, currently being operated in Arusha, Tanzania by Asian Vegetable Research and Development (AVRDC) and Ministry of Agriculture, Food Security and Cooperatives (MAFC), is a physical site where local postharvest/agro-industry research and extension personnel can meet and conduct practical adaptive research aimed at testing innovations under local conditions, and identify issues regarding practicality, costs/benefits, and potential returns. The PTSC provides demonstrations of those postharvest and food processing innovations determined to be feasible (both technically and financially), as well as comprehensive, hands-on training on improved postharvest practices, food safety, nutrition, and practical information for small-holders, traders, and women involved in postharvest/agro-industry activities and agribusinesses.

The PTSC is designed to be sustainably financed by operations (sponsored research projects, speciality training programs, and retail sales of goods and services) that generate revenue, which can be used to pay staff and for maintenance, utilities, and extension/training for groups who cannot afford to pay. This PTSC project was formally evaluated with US Agency for International Development (USAID)/Technical and Operational Performance Support (TOPS) Program funding by World Food Logistics Organization (WFLO) in 2014 and 2015, and was found to reach more than 1,500 local farmers, traders, food processors, and marketers via training programs provided in Tanzania by locally and internationally based postharvest trainers. One hundred percent of those who have participated in postharvest training programs in Tanzania reported being satisfied with their experiences, and making one or more changes in their practices that led to reduced losses of either fresh produce or processed products (Kitinoja and Barrett 2015). Of the 50 training participants evaluated via face to face interviews, 42 people were able to provide details on the costs and benefits of making changes in their traditional handling or processing practices for a range of fruits, vegetables, and herbs, and all of these same 42 reported increased earnings (WFLO 2010; Kitinoja and Barrett 2015).

5.4.4   FOOD LOSSES AT ENTRY POINT

One outcome of improved extension/advisory services on postharvest aspects should be improved quality supply management and reduces PHL. Food loss reduction has recently taken a priority place in the global political agenda. This has prompted actions in international forums and in countries. For instance, the Global Knowledge Initiative (GKI) consortium meeting in Nairobi in February 2015 reviewed the status of postharvest/agroprocessing/food storage programs, and recommended a systems approach to solving postharvest problems (GKI 2014). Other important FAO and World Bank work in extension can provide general background (HLPE 2004; FAO 2005; Qamar 2005; Swanson and Rajalahti 2010), but these reports have mostly provided frameworks for reviews/reforms focused on enhanced food production, and gave given very little guidance on how to provide extension services for postharvest technology and agro-industry development. The advice provided by Qamar (2005), however, applies to this sector as well:

> "The time is indeed ripe for policy-makers in developing countries to challenge and revisit the discipline of extension within a global context, so as to let the extension function be performed with excellence in line with the global challenges to their economies and especially to their agriculture sector. Cosmetic changes to the existing national extension systems will be of little benefit, as will be the repeated training of staff in stereotyped agricultural subjects."

By using standardized protocols for conducting assessments of food loss and training needs, researchers, extension workers, and clientele can work together to identify and solve local problems. FAO's SAVE FOOD initiative, for example, has been using a methodology for food loss assessment based on postharvest fish losses in small-scale fisheries (FAO 2011). The Commodity Systems Assessment Methodology (CSAM) is being promoted by the Postharvest Education Foundation (PEF) for the training of postharvest extension workers (PEF 2013; LaGra et al. 2016). Coordinated by the World Resource Institute (WRI), a food loss and waste accounting and reporting standard has been promoted for use by stakeholders, and can be downloaded at http://flwprotocol.org.

5.4.5   Return on Postharvest/Agro-Industry Investments

People will not adopt changes in postharvest practices or invest in new technologies for agro-industry development unless they are convinced of the positive return on investment, or ROI (Kader 2005; Kitinoja et al. 2011; Horton and Hoddinott 2015). Drawing from a report on postharvest investments in India and Sub-Saharan Africa (SSA) by WFLO (2010), Horton and Hoddinott (2015) illustrates some of the potential gains from the various technologies, including (1) curing of roots, tubers, and bulbs that leads to the return of a profit two-and-half times larger than the returns on non-adoption, and (2) cooling practices used for vegetables that can provide gains up to seven-and-half times higher than the initial costs.

Technologies like metal silos for grain storage can have high returns but are expensive to purchase and adopt (Gitonga et al. 2013). The major effect of adopting the metal silos in Kenya was the nearly total elimination of losses due to insect pests, saving farmers an average of 150–200 kg of grain, worth US$130. Metal silo adopters also spent about $5 less on storage insecticides. According to Gitonga et al. (2013) adopters were able to store their maize for 1.8–2.4 months longer, and to sell their surplus after five months at good prices, instead of having to sell right after the harvest when market prices tend to be low.

For rice milling using a village level milling machine of 40 kg/hour–300 kg/hour capacity, Horton and Hoddinott (2015) provided the following comparison:

- One-stage milling machine: consists of steel huller that accommodates husking and polishing; milling recovery is 50%–55%, which is extremely low, and head rice recovery less than 30%
- Two-stage milling machine: more sophisticated, with husker usually in the form of rubber roller and steel polisher; milling recovery above 60%

Clearly, the ROI in any given location will depend on the cost of the two-stage milling machine and the market prices for rice.

5.4.6   Impact of Current Laws to Provide Stimulus for
Private Entities to Enter the Service Sector

There are many examples of laws, regulations, and business requirements that make it more difficult for private sector organizations to provide extension/advisory services. Private sector consultants may require special business licenses, technical certifications, and liability insurance.

Agro-industry refers to the establishment of linkages between enterprises and supply chains for developing, transforming, and distributing specific inputs and products in the agriculture sector. Consequently, according to some experts, agro-industries are a subset of the agribusiness sector and will be affected by the laws and regulations imposed upon businesses in general. The PTSC described previously had difficulties in becoming established in Rwanda due to land use policies. In Tanzania, one of the Non-Governmental Organizations (NGOs) that was operating a PTSC was banned by the government from selling goods to the public, denying it as a major source of revenue that was intended to make the PTSC financially self-sufficient. Table 5.1 summarizes the effects of various policy-related factors on agro-industry development.

Decades ago, in response to food shortages, the government of India, under its Department of Agriculture and Cooperation, established the Agricultural Produce Marketing Committee (APMC),

TABLE 5.1

Examples of Policy-Based Stimulations and Constraints

Country	Policy	Positive	Negative
Rwanda	Land in or near city centers cannot be developed; land with water resources, drainage, etc. cannot be developed	Protects environment from over-development	Limits where businesses or agricultural training centers can be constructed
Tanzania	International NGOs cannot be licensed to sell goods or services		International NGOs active at the local level must find other mechanisms to provide agro-industry related goods/services
India	APMC Act prohibits transactions for all agricultural commodities outside of government regulated wholesale markets (see below)	Protects sellers from unscrupulous buyers	Limits options, slow transactions, requires transport to wholesale markets for physical exchange of goods
Most countries	Licensing, technical certification, liability insurance required for private contractors, consultants, technicians, businesses	Knowledge and skills are verified	Can limit the number of people who can afford to be involved
Most developing countries	Limited availability of credit at reasonable interest rates		Can limit the number of people who can afford to invest in improved postharvest handling technologies or new agroprocessing businesses

Source: Authors' analyses of desk study survey of key informants (2015).

The AMPC is governed by the APMC Act, which prohibits transactions for all agricultural commodities outside of the regulated wholesale markets. According to the Act, no direct marketing and direct procurement of agricultural produce can take place from farmers' fields, and the setting up of alternative markets is restricted. The APMC laws were created to ensure better prices for farmers through an open auction system, but instead it has led to the development of local vested interests and reduced marketing options for small farmers by limiting their access to emerging domestic retail and export markets. According to Kitinoja et al. (2011), the states of Punjab, Maharashtra, and Bihar, among others, have amended the APMC Act in order to permit private companies, such as RK Foodland, NDDB, Reliance Fresh, and ITC Limited to set up their networks to procure goods. In the Model Central APMC Act 2003, it is envisaged that eventually each state will be able to amend the Act to allow farmers to sell their commodities anywhere they want to do so, and the wholesale market taxation system is proposed to be removed. In 2017, an Internet-based platform for making these amendments was launched (known as e-NAM). These simple moves to direct marketing of perishable crops allows quicker handling, fewer delays between harvest and marketing, less unnecessary transport of crops without cooling to/from central markets, and no extra unloading/loading/stacking of produce to be weighed and taxed, all of which have helped to reduce losses.

5.4.7 Building a New Model for Postharvest Extension/Advisory Services

Key informants from six of the 20 countries included in the survey reported that they were not aware of any extension activity in the postharvest/agro-industry sector (Chile/VII Región del Maule only, Ethiopia, Lebanon, Pakistan, Tanzania, and Uganda), yet there are welcome signs from every corner of the globe that postharvest technology and agroprocessing is gaining increasing support and investment all along food value chains. In fact, disciplines beyond harvest have started to be included in extension/advisory services, including packing practices, cold chain management, storage, food processing, and food safety, but respondents found it difficult to estimate the number of extension/advisory service personnel who are working in these subfields of postharvest/ agro-industry in their countries.

Key informants were in favour of a more active interaction between producers, researchers, and extension agents. These three actors overlap in important ways—adaptive research is needed to test scientific knowledge in a local setting, working with producer and agro-industry organizations, and making any needed modifications before extending the information to potential users. Subject matter specialists (SMS) bridge the gap between the research establishment universities/research centers and the field level extension workers, who must be able to communicate complex information. Lead farmers, producer organizations, agroprocessing companies, programs that foster the development of new agroprocessed products, and trade association leaders can interact with extension workers, informing the extension service of current problems and bringing technical information back to their constituent groups. Modern times provide the opportunity to further link local extension efforts with a more global agro-industry extension network that can build a more efficient scheme (Figure 5.2). In this case, in implementing adaptive research, planning and implementation can be supported by a large "free to access" library of scientific documents and assessment tools. The technical advisory committee (TAC) includes all the key stakeholders in the technical, social, and agri-business sectors. National SMS staff requires regular training updates (known as in-service training) but they can also receive technical training and skills updates from outside organizations and via online e-courses, e-learning programs, and distance education graduate programs. The topics included in this field are

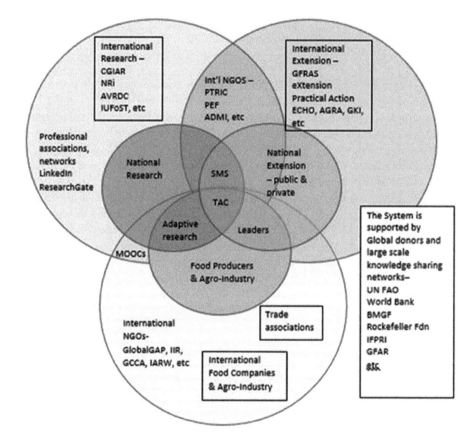

FIGURE 5.2   Model of efficient research extension/agro-industry system interactions.

extensive and dynamic: from technical practices used on the farm to enhance quality and nutrition, to postharvest handling practices used by various actors along the value chain, to food safety, food processing/manufacturing, packaging innovations, cold chain management, marketing and storage logistics, food retailing, and much more. Massive online open courses (MOOCs) are low cost opportunities for education, knowledge sharing, and capacity building, and are being offered by established agricultural universities and attended by thousands of participants from the food production and agro-industry sectors each year.

The classic steps of any extension program are to assess local or regional needs, gather resources, set measurable objectives, design extension education programs, implement, monitor, and evaluate. Governance of the extension/advisory services could be done via an advisory board or TAC, which would be at the helm and participate as key stakeholders in this cycle. The TAC, in conjunction with the administrators of the extension/advisory services, would be responsible for ensuring that the needed resources (financial, personnel, and support services) were sufficient to accomplish the organization's objectives and that resources were being used efficiently and effectively.

Enhancing the efficiency of national extension/advisory services and integrating postharvest capacity building efforts requires the consideration of the following factors:

- Most national extension/advisory services are still focused on increasing production of food crops, seafood, meats, and dairy products, while currently 30%–50% of the foods that are being produced is lost or wasted before consumption.
- National extension/advisory services do not need to stand alone and try to fill every agricultural and agro-industry or food science need-gap that is identified in their country using only local resources. A global support network has emerged in the past decade with a vast array of informational resources that are easy to access via the internet. It can provide low-cost program planning support, technical training, and monitoring and evaluation (M&E) advice for developing more efficient national extension/advisory services.
- Countries differ widely in the kinds of foods they produce, process, store, and eat. They also differ in climate, infrastructural base, and cultural/social capital, so it is important that, as an expanded extension/advisory service is being developed, each country focuses on the crops and food products that best suit their own national interests. This might include commodities that provide inclusiveness and sustainability to their society, or taking into account an urgent matter, such as crops for targeting specific nutrition issues or poor areas in the country.
- Gender relations and gender issues must be acknowledged in shaping the environment in which extension/advisory services operate. Gender relations should be part of a wider systems approach to improving agricultural and development goals.

The analysis of what has been done today in countries for integrating postharvest topics into a national system of advisory/extension services allows the recommendation of a model for enabling an appropriate environment. The model includes seven steps (Figure 5.3), with the following key objectives for more effective extension/advisory services:

- Meeting national development goals for food security, including access to adequate, safe, and nutritious foods
- Providing opportunities for enhancing rural livelihoods by creating new jobs in a growing postharvest/agro-industry sector
- Reducing food losses and waste in order to protect natural resources and ensure that the national food system is being developed in an environmentally sustainable way
- Ensuring any under-represented groups, such as women, smallholder producers, SMS, traders and youth involved in the food system have equitable access to productive resources and inputs, including agro-industry extension/advisory services

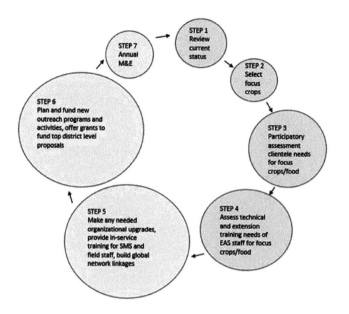

FIGURE 5.3 The seven steps model: guidelines for national planning of expanded extension/advisory services with an active postharvest/agro-industry focus.

A brief explanation of the steps:

1. National assessment of the status of existing extension/advisory services, and public and private sectors to identify the current versus needed levels of staffing, funding, and supporting operational resources (transport, tools, computers, mobile phones, and ICT services)
2. Investing in any missing tools, staff, and related operational support
3. Training needs assessments/analysis (TNA) for current and representative new clientele in the postharvest/agro-industry sector (e.g., packinghouses, transporters, food processors, traders, storage operators, direct marketers, exporters, wholesale and retail marketers, etc.)
4. TNA to identify the number and level of knowledge and skills of SMS in technical subject matter areas (production, postharvest technology, food science and agro-processing, and value chain management) and recommended extension education practices (group formation, training needs assessment, program planning, participatory teaching/learning methods, facilitation style extension/advisory approaches, and monitoring and evaluation practices), which will be required to meet the expressed needs of clientele
5. Investing in knowledge/skills upgrading for national SMS, and providing postharvest in-service training opportunities for field level staff, including building linkages to global resource networks
6. Knowledge/skills upgrading for private sector extension/advisory service providers (it may be possible to charge fees for annual certification)
7. Monitoring and evaluation in order to document outcomes and make improvements in each of the seven steps

5.5   CONCLUSION AND FINAL REMARKS

Analyses of the diverse outcomes and factors deriving from failures and successes in extension confirm there is no perfect plan on postharvest extension/advisory services that can be prescribed. Instead, a series of local assessments and setting of national priorities must be implemented over a period of time. Each country can make decisions regarding whether to support public sector extension/advisory services, or provide funding and backstop for the private sector based postharvest technology/agro-industry training and services, or some combination of the two.

There is little disagreement among experts and practitioners on whether functional extension/advisory services serve an important role in agricultural development, food security, and improved nutritional and socioeconomic status of rural populations. If national extension/advisory services choose to integrate the complex and dynamic sector of postharvest/agro-industry, it is likely that the size of the existing services will need to double. For each extension/advisory service staff member assigned to agricultural production for a key crop or food group, another staff member should be assigned to provide advice and training related to the postharvest aspects of the related value chains. This may seem to be a daunting task, but the investment would be worthwhile and ultimately profitable, since economists from International Food Policy Research Institute (IFPRI) have estimated that the overall cost of approximately halving PHL in the developing world would be US$239 billion over the next 15 years, and would generate benefits worth more than $3 trillion, or $13 of social benefits for every dollar spent (Horton and Hoddinott 2015). For smaller countries, this kind of payback could be planned and experienced as an investment of $1 million for an expected return of $13 million, which falls well within the current range of donor-sponsored projects.

For those policymakers and global development planning bodies that take a longer-term view and are convinced of the need to reduce food losses, the target level of investment proposed by IFPRI is within reach, at approximately $16 billion per annum. With one out of every four kilocalories produced on the planet currently not being consumed, an estimated $32 billion/year worth of foods being thrown away in China alone (Lipinski et al. 2013), and production resources becoming scarcer or under pressure from climate changes, investing in expanded and enhanced extension/advisory systems make economic, as well as environmental, sense.

The most likely threat to efforts toward modernizing existing production-oriented extension/advisory services is limited funds for making the required systematic assessments and upgrades. A financial commitment will be required for several major start-up activities and full implementation of the 7 Steps if policymakers and extension service administrators plan to successfully implement the guidelines.

The outlook regarding opportunities is limited only by imagination and financial commitment. The realization that any individual country is not standing alone in their quest to expand their national extension/advisory services but can become part of a vast and growing global network of technical information and education on agricultural production, postharvest technology and agro-processing, extension/advisory service support and educational resources, should make the desired expansion relatively straightforward to plan and accomplish.

ACKNOWLEDGMENTS

Alejandra Safa (UN FAO ESP)
Tomoko Kato (UN FAO ESN)
Rome, Italy

REFERENCES

Cho, K. M. (2002). Training needs of agricultural extension agents in Myanmar. In: *Proceedings of the 18th Annual Conference*, Durban, South Africa, AIAEE 2002, pp. 72–80.

FAO. (2005). *Enhancing Coordination among AKIS/RD Actors: An Analytical and Comparative Review of Country Studies on Agricultural Knowledge and Information Systems for Rural Development (AKIS/RD)*. By W.M. Rivera, M.K. Qamar and H.K. Mwandemere. Rome, Italy.

FAO. (2011). Post-harvest fish loss assessment in small-scale fisheries. FAO Fisheries and Aquaculture Technical Paper 559.

Farnworth, C. R. and Colverson, K. E. (2015). Building a gender-transformative extension and advisory facilitation system in Sub-Saharan Africa. *Journal of Gender, Agriculture and Food Security*, 1(1), 20–39.

Gitonga, Z. M., De Groote, H., Kassie, M. and Tefera, T. (2013). Impact of metal silos on households' maize storage, storage losses and food security: An application of a propensity score matching. *Food Policy*, 43, 44–55.

GKI. (2014). Reducing food waste and spoilage: Assessing resources needed and available to reduce postharvest food loss in Africa. Global Knowledge Initiative, June 2014. Retrieved from http://postharvest.org/Rockefeller%20Foundation%20Food%20Waste%20and%20Spoilage%20initiative%20Resource%20Assessment_GKI.pdf

Gustavsson, J., Cederberg, C., Sonesson, U. and Emanuelsson, A. (2013). SIK report No. 857 The methodology of the FAO study: "Global food losses and food waste - extent, causes and prevention"- FAO, 2011. The Swedish Institute for Food and Biotechnology. Retrieved from http://www.diva-portal.se/smash/get/diva2:944159/FULLTEXT01.pdf

HLPE. (2014). Food losses and waste in the context of sustainable food systems UN FAO HLPE on food security and nutrition. July 2014. p. 117. Retrieved from http://www.fao.org/fileadmin/user_upload/hlpe/hlpe_documents/HLPE_Reports/HLPE-Report-8_EN.pdf

Horton, S. and Hoddinott, J. (2015). Benefits and costs of the food security and nutrition targets for the post-2015 development agenda—Post-2015 Consensus. IFPRI: Copenhagen Consensus Center Working Paper.

Kader, A. A. (2005). Increasing food availability by reducing postharvest losses of fresh produce. *Acta Horticulturae*, 682, 2169–2175.

Kitinoja, L. and Al Hassan, H. Y. (2012). Identification of appropriate postharvest technologies for improving market access and incomes for small horticultural farmers in Sub-Saharan Africa and South Asia—part 1. postharvest losses and quality assessments. *Acta Horticulturae* (18th IHC 2010), 934, 31–40.

Kitinoja, L. and Barrett, D. M. (2015). Extension of small-scale postharvest horticulture technologies—A model training and services center. *Agriculture*, 5(3), 441–455. Retrieved from http://www.mdpi.com/2077-0472/5/3/441/pdf

Kitinoja, L., Saran, S., Roy, S. K. and Kader, A. A. (2011). Postharvest technology for developing countries: Challenges and opportunities in research, outreach and advocacy. *Journal of the Science of Food and Agriculture*, 91(4), 597–603.

LaGra, J., Kitinoja, L. and Alpizar, K. (2016). *Commodity Systems Assessment Methodology for Value Chain Problem and Project Identification: A First Step in Food Loss Reduction*. Costa Rica: IICA. Retrieved from http://repiica.iica.int/docs/B4232i/B4232i.pdf

Lipinski, B., Hanson, C., Lomax, J., Kitinoja, L., Waite, R. and Searchinger, T. (2013). Installment 2 of creating a sustainable food future: Reducing food loss and waste. World Resources Institute Working Paper: June 2013. Retrieved from http://pdf.wri.org/reducing_food_loss_and_waste.pdf

Manfre, C. and Nordehn, C. (2013). Exploring the promise of information and communication technologies for women farmers in Kenya. Cultural Practice, LLC MEAS Case Study # 4, August 2013. Retrieved from http://pdf.usaid.gov/pdf_docs/PA00KVFN.pdf

Manfre, C. and Sebstad, J. (2011). FIELD Report 12: Behavior change perspectives on gender and value chain development: Framework for analysis and implementation. Field-support LWA. Washington, DC: ACDI/VOCA and FHI 360 Retrieved from https://microlinks.org/sites/microlinks/files/resource/files/FIELD%20Report%20No%2012_Gender%20and%20Behavior%20Change%20Framework.pdf

Mohamed, I. E., Gamon, J. A. and Trede, L. D. (1995). Third world agricultural extension organizations - obligations toward the education needs of rural people: A national survey. *Journal of International Agricultural and Extension Education*, 2(2), 95. Retrieved from https://www.aiaee.org/attachments/article/426/Mohamed-Vol-2.2–2.pdf

PEF. (2013). Gathering data to address postharvest loss challenges: Commodity systems assessment methodology. White Paper No. 13-02. La Pine, OR: The Postharvest Education Foundation. pp. 8. Retrieved from http://postharvest.org/CSAM%20Gathering%20data%20on%20Postharvest%20loss%20challenges.pdf

Qamar, M. K. (2005). *Modernizing National Agricultural Extension Systems: A Practical Guide for Policymakers of Developing Countries*. Rome, Italy: FAO Research, Extension and Training Division. Retrieved from http://www.fao.org/uploads/media/modernizing%20national.pdf

Quisumbing, A. R. and Pandolfelli, L. (2009). Promising approaches to address the needs of poor female farmers—Resources, constraints and interventions. IFPRI Discussion Paper 00882. Washington, DC: International Food Policy Research Institute. Retrieved from http://ebrary.ifpri.org/utils/getfile/collection/p15738coll2/id/27075/filename/27076.pdf

Swanson, B. E. and Rajalahti, R. (2010). Strengthening agricultural extension and advisory systems: Procedures for assessing, transforming, and evaluating extension systems. World Bank Agriculture and Rural Development Discussion Paper 45. 187p. Retrieved from http://siteresources.worldbank.org/INTARD/Resources/Stren_combined_web.pdf

WFLO. (2010). Identification of appropriate postharvest technologies for improving market access and incomes for small horticultural farmers in Sub-Saharan Africa and South Asia. WFLO Grant Final Report to the Bill & Melinda Gates Foundation, March 2010. p. 318.

Wongtschowski, M., Belt, J., Heemskerk, W. and Kahan, D. (Eds.). (2013). *The Business of Agricultural Business Services: Working with Smallholders in Africa.* Royal Tropical Institute, Amsterdam; Food and Agriculture Organization of the United Nations, Rome, Italy; and Agri-ProFocus, Arnhem. Retrieved from https://www.kit.nl/sed/wp-content/uploads/publications/2080_the_business_of_agricultural_business_services.pdf

6 World Food Preservation Center® LLC

Charles L. Wilson

CONTENTS

6.1 INTRODUCTION AND MISSION

On its present course, the world is not going to produce food fast enough to keep up with a rapidly increasing global population that is expected to reach 9.6 billion people by 2050 (Ray et al. 2013). One-third of the food that we already produce worldwide is lost between the time that it is harvested and the time that it would be consumed (FAO 2011). Subsequently, it is imperative that we save substantially more of the food that we already produce if we are to avoid accelerated world hunger in the future.

A major reason that we are suffering major food losses after harvest is that we have invested so few of our agricultural resources in the postharvest preservation of food (5%), as compared to our investment in food production (95%) (Kader and Rolle 2004). This meager investment in the postharvest preservation of food has left us with postharvest "skill gaps" and "technology gaps," particularly in developing countries. The mission of the World Food Preservation Center® LLC (WFPC) is the filling of these "postharvest intellectual gaps" in developing countries (http://www.worldfoodpreservationcenter.com/index.html) (Kader, 2013; Wilson, 2013a,b).

The WFPC is accomplishing its mission by (1) promoting the postharvest education (MS and PhD) of young student/scientists from developing countries in the latest technologies for the postharvest preservation of food, (2) the development of postharvest curricula/texts for the secondary education of students in developing countries), (3) the organization of continent-wide postharvest congresses and exhibitions, (4) the publication of a postharvest reference/text book targeted toward the needs of developing countries, and (5) the organization of a Global Mycotoxin Alliance to focus the world's resources on reducing mycotoxin contamination of food in developing countries (http://www.worldfoodpreservationcenter.com/index.html).

6.2   HISTORY

On a LinkedIn Postharvest Training discussion group administered by Dr. Lisa Kitinoja, Founder of The Postharvest Education Foundation (PEF), Dr. Charles L. Wilson entered the following post April 24, 2012, "How about a World Food Preservation Center® LLC that would train young scientists in developing countries in advanced postharvest technologies and develop new technologies such as solar refrigeration? (Wilson, 2012)"

The response to this post was immediate and overwhelmingly positive. Young student/scientists from over 40 different developing countries expressed their support for this idea and indicated their willingness to participate. As a result, Dr. Wilson presented the idea of a WFPC to Professor Hongyin Zhang and Dean Haile Ma at Jiangsu University in China, who agreed to their university becoming a "sister university." The president of Jiangsu University, Dr. Juan Shouqi, provides a full PhD postharvest scholarship to the WFPC program each year.

The first PhD graduate from the WFPC program was Dr. Gustav Mahunu from Ghana who received his PhD degree under the supervision of Professor Zhang at Jiangsu University. Dr. Mahunu has now returned to the University for Development Studies in Ghana, which has also joined the WFPC as a sister university, and has established his own independent postharvest program.

6.3   PRESENT STATUS

The WFPC now comprises 28 major agricultural research universities and three major agricultural research institutes on six continents. Sister universities of the WFPC include: Jiangsu University (China), University of Florida (United States), Cranfield University (United Kingdom), University of Sydney (Australia), University of Ghana, FUNAAB (Nigeria), University of Nairobi (Kenya), University of Chile, UFRGS Brazil , KU Leuven (Belgium), Haramaya University (Ethiopia), Allahabad University (India), Tamil Nadu Agricultural University (India), University of Lisbon (Portugal), University of Hawaii (United States), University of Agriculture Faisalabad (Pakistan), University of Guelph (Canada), University of Manitoba (Canada), Jimma University (Ethiopia), Universidad Politecnica de Cartagena (Spain), La Plata University (Argentina), University of KwaZulu of Natal (South Africa), Integral University (India), University for Development Studies (Ghana), University of Rwanda, Ghent University (Belgium), and Foggia University (Italy) (Figure 6.1).

The Volcani Center in Israel (equivalent to the United States Department of Agriculture [USDA]); INRA in Morocco; and the Philippine Center for Postharvest Development and Mechanization have joined the WFPC as "sister institutes."

FIGURE 6.1 Sister universities and institutes of the WFPC.

6.4 POSTHARVEST EDUCATIONAL PROGRAMS

6.4.1 PhD and MS Postharvest Scholarships

The WFPC is promoting the PhD and MS postharvest education of young student/scientists from developing countries at its sister universities. In order to solicit funds for these scholarships the non-profit 501(c)(3), World Food Preservation Education Foundation (http://www.foodpreservationfoundation.org/index.html) has been formed.

6.4.2 Recommended Postharvest Research Programs

Based on the recommendations of its advisors, the WFPC recommends the following structure and topics for the development of university postharvest programs in developing countries.

6.4.2.1 Engineering Department

The Engineering Department of the WFPC will conduct research on:

1. Low cost refrigeration (solar, evapo-cooling, etc.) adaptable for the storage, transport, and marketing of food that is suitable for developing countries
2. Drying and long-term storage technology for the preservation of food that is suitable for developing countries
3. Physical postharvest treatments that extend the shelf life of harvested commodities such as Ultraviolet C, gamma irradiation, etc.
4. Low cost harvesting and processing equipment allowing increased efficiency and minimizing postharvest damage in developing countries
5. Creating innovative storage environments to extend the storage life of fruits, vegetables, and grains

6. Developing cost-effective cool chain management technologies for use with foods that need to be cooled but that are not typically kept in long-term storage at very low temperatures (fresh or frozen)
7. Adapting food transportation vehicles so as to minimize vibration and abrasion damage during food transport
8. Developing simple, non-destructive harvesting tools for fruits, vegetables, and grains
9. Farm processing technology to reduce postharvest losses (PHL)
10. Development of low cost transportation systems for harvested commodities
11. Nanotechnology for reducing microbial contamination on surfaces and strengthening and reducing gas exchange of plastic bags and containers

6.4.2.2   Postharvest Physiology Department

The Postharvest Physiology Department of the WFPC will conduct research on:

1. Manipulating preharvest physiology to impact on postharvest physiology to extend storage shelf life
2. Determining the effects of modified and controlled atmosphere storage on shelf life
3. Studying the impacts of abscission on postharvest quality
4. Development of innovative ripening and senescence indices for fruits, vegetables, and grains
5. Development of texture indices (outer and inner) and color assessment indices (inner and outer)
6. Interventions in the ripening and senescence process (ethylene inhibition, etc.)
7. Manipulating postharvest biochemical changes to enhance nutritional quality
8. Manipulating the physiology of postharvest storage of pome fruit, stone fruit, subtropical fruit, vegetables, ornamentals, grain, and milk to extend storage life
9. Reducing postharvest storage damage to fruits, vegetables, and grains, particularly low temperature damage to tropical produce
10. Reducing PHL by managing water losses in harvested and stored commodities.
11. Defining the impact of degree days and storage environments on the postharvest behavior of native foods

6.4.2.3   Food Processing Department

The Food Processing Department of the WFPC will conduct research on technologies that will extend the shelf life of harvested food and add value to it (e.g., gari from cassava), as well as develop technologies that assure food safety, such as:

1. Innovative technologies that add value to food and expand market potential (e.g., cassava chips, gari, dried fruits, and fermented foods)
2. Enhancing nutrients and nutraceuticals in harvested commodities through physical and biological means
3. Phytosanitation practices that reduce microbial and toxin (aflatoxin) contamination
4. Simple detection technologies for microbial foodborne pathogens and aflatoxins
5. Determining the impact of the application of postharvest technologies on the nutritional value of food
6. Determining the impact of the microbiome in the food chain on PHL

6.4.2.4   Food Packaging and Storage Department

The Food Packaging and Storage Department of the WFPC will conduct research on:

1. Active packaging for harvested commodities, including scrubbers (oxygen, carbon dioxide, ethylene, etc.), volatile antimicrobials, and nano-antimicrobial surface coatings
2. Intelligent packaging with radio-frequency identification (RFID) sensors and activators

3. Modified polymer films for modified atmosphere packaging (MAP) and controlled atmosphere packaging (CAP)
4. Carbon dioxide generators for food storage
5. Hermetically sealed containers for harvested food
6. Protective containers for food transport
7. Silos and storage buildings
8. Edible antimicrobial and MAP coatings

6.4.2.5 Postharvest Pest Control Department

The Postharvest Pest Control Department will conduct research on the control of postharvest insects and disease pests that do not require pesticides, are biologically based, and that are adaptable for developing countries

1. Physical (radiation, heat, etc.) and biological control (microbial antagonists) of postharvest diseases of fruits and vegetables
2. Physical and biological control of postharvest diseases of grains
3. Physical and biological control of microorganisms in meat and dairy
4. Management of preharvest conditions to reduce postharvest pests
5. Harvesting technologies to reduce postharvest pests
6. Natural pesticides (pesticidal plant extracts, pheromones, etc.)
7. Integrated pest management (IPM)
8. Natural enemies of pests
9. Phytosanitation and pest control

6.4.2.6 Postharvest Genetics Department

The Postharvest Genetics Department of the WFPC will conduct research on:

1. Identification of genes and varieties with extended storage and transport resistance to loss in harvested fruits, vegetables, and grains
2. Selection and breeding of plant varieties with greater PHL resistance
3. Developing commodities with reduced chilling sensitivities
4. Identification of genes related to resistance to postharvest pests
5. Identification of genes related to cold hardiness
6. Identification of genes related to ethylene sensitivity
7. Identification of genes related to skin permeability

6.4.2.7 Postharvest Extension Department

The Postharvest Extension Department will conduct research, training, and outreach activities aimed at increasing awareness of the need for and adoption of cost-effective, improved food preservation practices and postharvest technologies suitable for developing countries, such as:

1. Training on assessing the physical volume and economic value of PHL and quality changes of foods (protocols, data collection methods, and tools for measuring quality, loss, and waste)
2. Offering "training of trainers" programs on the extension of appropriate food preservation technologies via formal (classroom) and informal (workshops, short courses, study tours, e-learning) methods
3. Managing a Virtual WFPC network via an online interactive website for mentoring, educating, and supporting WFPC students and young professionals worldwide
4. Research on models for learning and extension methods for dissemination of knowledge and skills in food preservation, in order to improve outreach efforts and impact of WFPC extension programs

6.4.3 Secondary/Vocational School Curricula/Texts

According to Irina Bokova, Director General of the United Nations Educational, Scientific, and Cultural Organization (UNESCO), "There can be no escape from poverty without a vast expansion of secondary education. This is a minimum entitlement for equipping youth with the knowledge and skills they need to secure decent livelihoods in today's globalized world." Yet, two-thirds of African children are effectively locked out of secondary school, according to a new UN report that cites secondary education as one of the next great development challenges facing many of the world's poorest countries.

The WFPC recognizes that not only is it important to dramatically increase support for secondary education in developing countries, we must also target and make relevant the curricula that students attending these schools receive. The WFPC, through its network, is developing food loss/safety/nutrition curricula for secondary schools in Africa, Latin American, and Asia (https://www.linkedin.com/pulse/targeted-investments-agricultural-education-countries-wilson/).

The food curricula being developed to educate young people in developing countries in the latest technologies for preserving food and keeping food safe and nutritious will serve multiple purposes. They will introduce young students to the ideas that (1) agriculture involves not only the production of food but also the preservation of food, (2) agriculture is not just hard labor, but a science and profession, (3) there are job opportunities in agriculture after food is harvested, (4) food contamination is a threat to the safety of their family and animals, and (5) if food is not properly handled it can rapidly lose its nutritional value.

6.5 CONTINENT-WIDE POSTHARVEST CONGRESSES

Numerous independent initiatives have been launched globally to combat postharvest food losses in developing countries. In order to coordinate these activities and forge new relationships, the WFPC has organized three continent-wide postharvest congresses and exhibitions in Africa, Latin American, and Asia.

The First All Africa Postharvest Congress and Exhibition was held in Nairobi, Kenya March 28–31, 2017 with over 600 in attendance from 40 countries (http://www.worldfoodpreservation-center.com/1st-all-africa-postharvest-congress-and-exhibition.html). The First All Latin America/Caribbean and All Asian Postharvest Congress and Exhibition (Bento Gonçalves, Brazil), and the First All Asian Postharvest Congress and Exhibition (Lucknow, India) will both be held in 2019.

6.6 WFPC® BOOK SERIES

The WFPC, in conjunction with CRC Press (Taylor & Francis), is publishing the WFPC Book Series, which contains advanced postharvest reference books and textbooks directed toward postharvest technologies and methodologies useful to developing countries, such as solar refrigeration, postharvest biological control, hermetic storage, and solar drying. Eleven books have been proposed, and four books are now under contract.

6.7 FUTURE PLANS

6.7.1 Organization of a Global Mycotoxin Alliance

Mycotoxins are recognized as causing major health problems in people and animals in developing countries. They are cancer causing and contribute to childhood stunting. In conjunction with its sister university, Ghent University, and BioIntelliPro, the WFPC is proposing a Global Mycotoxin Alliance that will bring together the world's leading mycotoxin experts to make available advanced detection and biocontrol technologies to developing countries for the detection and control of mycotoxins. This proposal is centred around Ghent University's preeminence in the detection of mycotoxins and the University of Arizona's expertise in the biocontrol of mycotoxins. The University of Ghent has already initiated the first phase of this initiative, which educates young scientists in Africa in the latest technologies for the detection of mycotoxins (http://mytoxsouth.org/). The WFPC is proposing extending this program to Latin America and Asia.

6.7.2 VIRTUAL WFPC

The WFPC is planning a Virtual WFPC, which will be an online database containing all the world's knowledge on the postharvest preservation of food in developing countries, with access portals for researchers, students, agricultural administrators, businesses, and farmers.

6.8 CONCLUSION

The major response toward meeting the world's pending food shortage crisis has been to mount another "Green Revolution" and produce more food. Yet, we know that we are not going to be able to produce food fast enough to feed the world's rapidly exploding population. So, what are our alternatives?

Since we know that we are losing enough food globally every year to feed two billion hungry people, we must save substantially more of this lost food in order to keep world hunger from escalating. Because of our underinvestment in postharvest technology and education, we find ourselves with a major postharvest "skill gap" and "technology gap," particularly in developing countries that suffer the most from postharvest food losses. The mission of the WFPC is to raise awareness of the importance of postharvest food losses in our fight against world hunger, and to launch initiatives that enhance the "postharvest intellectual capital" of developing countries.

REFERENCES

FAO. (2011). *Global Food Losses and Food Waste – Extent, Causes and Prevention*. Rome, Italy: FAO. Retrieved from http://www.fao.org/docrep/014/mb060e/mb060e00.pdf

Kader, A. A. (2013). Opportunities for international collaboration in postharvest education and extension activities. *Acta Horticulturae*, 1012, 1363–1370.

Kader, A. A., and Rolle, R. S. (2004). *The Role of Post-Harvest Management in Assuring the Quality and Safety of Horticultural Produce* (Vol. 152). Rome, Italy: Food and Agriculture Organization of the United Nations.

Ray, D. K., Mueller, N. D., West, P. C., and Foley, J. A. (2013). Yield trends are insufficient to double global crop production by 2050. *PLoS One*, 8(6), e66428.

Wilson, C. L. (2012). How about a World Food Preservation Center® LLC that would train young scientists in developing countries in advanced postharvest technologies and develop new technologies such as solar refrigeration? [LinkedIn]. Retrieved from http://www.linkedin.com/groupAnswers?viewQuestionAndAnswers=discussionID=110145279gid=3770124commentID=80429933trk=view_discut=36TxbG6mIC15g1

Wilson, C. L. (2013a). Establishment of a World Food Preservation Center® LLC. *Agriculture and Food Security*, 2 (1), 1. Retrieved from https://agricultureandfoodsecurity.biomedcentral.com/articles/10.1186/2048-7010-2-1

Wilson, C. L. (2013b). Establishment of a World Food Preservation Center® LLC: Concept, need, and opportunity. *Journal of Postharvest Technology*, 1 (1), 1–7. Retrieved from http://jpht.info/index.php/jpht/article/view/17423/8939

7 Reducing Postharvest Loss and Waste in Fruits and Vegetables
Amity's Initiatives

Sunil Saran and Neeru Dubey

CONTENTS

7.1 INTRODUCTION AND IMPORTANCE

The present nutritional status of India would appear enigmatic. That the second-largest producer of food should, at the same time, have the second-highest undernourished population in the world is alarming, and calls for unravelling the cause(s) behind the paradox and taking remedial steps to mitigate the maladies. A look at the present food scenario in India would reveal that the country is not only self-sufficient in food but also export agriproducts. It is one of the top producers of cereals (wheat and rice), pulses, fruits, vegetables, milk, meat, and marine fish. In 2016–2017, production of fruits and vegetables increased to reach nearly 90% of the total production of 300 million MT (Metric Tonnes) of the country's horticultural produce, with fruits at 93 million and vegetables at 176 million (Indian Horticulture Database, 2016–2017, National Horticulture Board). Globally, India ranks first in production of banana (22.04%), papaya (40.74%), and mango (32.65%). It is also the largest producer of okra in vegetables, and ranks second in production of potato, onion, cauliflower, eggplant, and cabbage (Indian Horticulture Database, 2016–2017, National Horticulture Board). The stark reality of hunger and undernourishment in India, in the midst of self-sufficiency, comes to the fore when one looks at the Global Hunger Index (GHI) 2017 report, released by International Food Policy Research Institute (IFPRI) in October 2017. According to this report, India ranked 100th among 119 countries surveyed on GHI, with a score of 31.4, and placed in the "serious" category. In the four parameters that are taken as indicators of GHI, India's performance has been depressing, with the proportion of undernourished in the population at 14.5%, prevalence of wasting/thinness (low weight-for-height) in children under 5 years at 21.0%, prevalence of stunting (low height-for-age) in children under 5 years at 38.4%, and under-five mortality rates at 4.8%. India's own National Family Health Survey-4 (2015–2016) was carried out by Ministry of Health and Family Welfare; data on the prevalence of various indicators were, more or less, similar. Another serious note was struck by the revelation that 53% of all women in India in the age group of 15–49 were anemic (had a deficiency of blood hemoglobin). The overall picture related to hunger, hidden hunger, and other nutritional problems revealed that India was one of the worst performers in Asia, better than only Afghanistan and Pakistan. To address the issues of undernutrition, stunting, anemia, and low birth weight, the Union Cabinet has approved setting up of National Nutrition Mission (NNM), which will work in three phases from 2017–2018 to 2019–2020, and cover all districts of the country.

The foregoing update on India's poor nutritional status highlights the fact that there is a major gap between the food produced and the food consumed. According to data published by the Ministry of Food Processing Industries on August 9, 2016, harvest and postharvest loss (PHL) of India's major agricultural produce is estimated at US$13 billion. The present scenario, related to food loss and food waste, reinforces the importance of Amity University's initiatives in postharvest technology, as postharvest technology can serve as an important resource for combating hunger and enhancing nutritional status.

Inspired and motivated by Adel A. Kader and Lisa Kitinoja during their visit to India as resource persons in a series of workshops conducted by US Trade and Development Agency (USTDA), Amity University at Noida launched a full-fledged center, Amity International Centre for Postharvest Technology and Cold Chain Management (AICPHT and CCM)—in the year 2008. The main focus of this center was to provide a strong base for study and research in postharvest and cool chain management, the two areas of horticulture imbued with potentials of revolutionizing horticulture and horti-business in India, and to develop trained personnel to fill the huge demand in developing countries for tackling the problems emerging from postharvest horticultural loss and waste.

7.2 RESEARCH, TEACHING, AND EDUCATION INITIATIVES BY AICPHT AND CCM

AICPHT and CCM has a mandate to spread awareness about postharvest management, and has been working in this area for the last 10 years. So far, it has worked on 10 prestigious projects. The initiatives have been grouped into three different heads—networking and outreach, extension, and technology transfer and teaching—all of which are detailed below.

7.2.1 Networking and Outreach Initiatives

7.2.1.1 Study on Postharvest Losses and Development of Suitable Technologies to Reduce Them

A study in collaboration with World Food Logistics Organization (WFLO), under a project titled "Identification of appropriate postharvest technologies for improving market access and incomes for small farmers in Sub-Saharan Africa and South Asia," funded by Bill and Melinda Gates Foundation, 2009, entailed compilation of Commodity System Assessment Methodology (CSAM) reports for four fruits and ten vegetables for assessment of losses at various levels, from production to postharvest stages. Systematic assessment and characterization of PHL in the selected 14 key horticultural crops were carried out by using field-based measurements along the supply chain, from farm to retail, using a modified CSAM. A set of technologies were transferred to the farmer's field and cost-benefit ratios were calculated. The details of some of these technologies transferred were as follows.

7.2.1.1.1 Protection of Harvested Produce

A suitable portable shade was provided near the farmer's field so that the harvested produce could be shifted to the shade immediately after harvest (Figure 7.1). The portable shade was made by spreading a polynet over a metallic supporting structure, which prevented direct exposure of the fresh produce to the sun. On-farm operations like gathering/collection of produce, sorting, grading, packing, etc., could be undertaken under this shed instead of the open sky. Shade greatly reduces the temperature and helps in the removal of field heat, which directly affects the rate of respiration and ripening of any fresh produce that is being handled outdoors (Kitinoja et al. 2012).

7.2.1.1.2 Modified Wrapping and Cushioning Material

7.2.1.1.2.1 Stretch/Cling Film
Cling films made of polyethene or polypropylene, ideal for wrapping curds of cauliflower, were used to help keep the produce hygienic and prevent mishandling by the buyers, thus helping the curds retain their original whiteness without any trace of spoilage resulting from multiple handlings. Cling films were also used for covering bottle gourds, which also suffer disfiguration caused by buyers testing the freshness of the produce by digging thumbnails into the pulp. In cauliflower, the cost of transport is significantly reduced if only cling film wrapped curds are placed in Corrugated fiber board (CFB) lined crates and transported. The stalks and leaves, which together weigh around 60% of the cauliflower, are removed on the farm, where they can be sold. The core of the cauliflower stalk is cooked and eaten as a delicacy, or it can be pickled. The leaves, which are rich in β-carotene, can be blanched, dried, and powdered, and used as a food supplement.

7.2.1.1.2.2 Crate Liners
Rough edges of the plastic crates can cause bruising and damage to the produce. Handlers generally use newspaper, which carries harmful printing ink that can leach into the produce. Ventilated CFB liners (recyclable) are a most suitable cushioning material. These liners are made of three-fluted corrugated fiberboard that resists any bruising or other storage damage in fruits and vegetables. The cost of 100 sets of five CFB liners used is less than US$15.

7.2.1.1.3 Low Cost Storage Technologies

7.2.1.1.3.1 Zero Energy Cool Chamber
The Pusa zero energy cool chamber (Pusa ZECC), which works on the principle of direct evaporative cooling, is ideal for low cost short duration storage of fresh produce (Roy 1985). The importance of this storage facility lies in the fact that it does

FIGURE 7.1   Portable shade.

TABLE 7.1
Cost of Cool Chamber (100 kg Capacity)

S. No.	Particulars	Amount (US$)
1.	Bricks (quantity 400)	31.25
2.	River bed sand (10 bags)	7.81
3.	Bamboo, straws, Khaskhas, etc., for top cover	7.81
4.	Corrugated tin shed	23.43
5.	Water tank, polythene sheets, drip pipe	15.62
6.	Labor	7.81
	Total	93.75

not require any electricity or energy to operate, and all the materials used in the construction of a cool chamber are easily available at village level and at low costs. Even an untrained person can construct it with one-day training, as it does not require any specialized skills. The materials used in the chamber, like brick and sand, are reusable. The cool chamber needs constant water saturation, either by a sprinkling of water once in the morning and once in the evening, or by fixing a drip system. The cool chamber can reduce the temperature by 10°C–15°C in comparison to the outside temperature, and maintains the high humidity of about 90%–95% (Dubey et al. 2013).

The farmers can use ZECC for on-farm pre-cooling, or to store produce for a few days so that they are not forced to make a distress sale at low prices (Table 7.1). Storage in ZECC successfully reduces PHL and pays for itself in a short time.

7.2.1.2 International Network

In 2012 AICPHT and CCM participated as a network partner in a project entitled "International network on Preserving safety and nutrition of indigenous fruits and their derivatives," funded by The Leverhulme Trust of the United Kingdom. The network partners were: Centre for Underutilized Crops; University of Southampton, UK; Qualisud Department, CIRAD, France; Amity University, India; Horticulture Research Centre, BARI, Bangladesh; Faculty of Food Science and Technology, Nong Lam University, Vietnam; Department of Food Technology and Chemical Engineering, Institute of Technology, Cambodia; and World Agroforestry Centre, University of Peradeniya, Sri Lanka.

The network objectives were to promote knowledge transfer and foster cooperation, development of human resources between partner countries and different types of organizations, and to promote utilization of underutilized fruits. Five workshops were organized under the project, covering the following areas: knowledge mapping, postharvest and cold chain technology, chemical and nutritional characterization, food safety standards, uptake, and dissemination. The workshop objectives were:

1. To provide training in the latest techniques in postharvest technology and cold chain management of indigenous fruits and their products
2. To strengthen and extend the existing network
3. To identify research gaps in the area of postharvest technology and cold chain management of indigenous fruits
4. To initiate development of research proposals in the area of postharvest technology and cold chain management of indigenous fruits

7.2.2 Research Initiatives

7.2.2.1 Cool Rooms and Cool Transport for Small-Scale Farmers

The project, funded by Horticulture Collaborative Research Support Project-United States Agency for International Development (HortCRSP-USAID), was carried out in collaboration with University of California, Davis (United States) and partnering countries Honduras and Uganda. CoolBot is a device (developed by Mr. Ron Khosla, www.storeitcold.com) that, when connected to a room air conditioner in an insulated room, can serve as a low cost substitute to refrigerated storage. CoolBot has a microcontroller brain with three sensors that sense the room temperature, and control the air-conditioner thermostat and ice formation on the evaporative coils, and which allow it to work even during high summer temperatures and when the cool room door is frequently opened. At Amity University, the walls of the room were insulated by clay and rice husk (1 m^3), and by extruded poly-styrene in the roof and floor, whereas the door was polyurethane foam (PUF) insulated. Shelf life of different fruits and vegetables were evaluated in the CoolBot cool room, and it was found that well-maintained low storage temperatures, according to the stored commodity, extended the com-modity's shelf-life and kept it marketable and consumable for a long time (Figure 7.2).

AICPHT and CCM was contracted by AgriSystem International (ASI), under its Sunhara proj-ect, to construct a CoolBot-fitted cool room in horti-retail market in Sultanpur, Uttar Pradesh. The constructed cool room successfully maintained the set temperatures and relative humidity, and directly benefitted the farmers and retailers by allowing them to store and market their produce over an extended period of time, rather than engaging in distress sales when prices are low, or dump-ing their produce in the field or on the roads (Dubey and Raman 2016). This room can serve as a replacement for high-cost cold storages for farmer groups and wholesale/retail markets.

The farmers and retailers used the CoolBot cool room at the retailer's *mandi* (horticulture mar-ket) in Sultanpur, Uttar Pradesh for short duration storage. The storage cost is charged on a per crate basis, and the retailers sell the produce as per the demand. This has resulted in loss reduction at *mandi* level to just 1%–2%. The retailers are earning around US$350–$400 per month with a period of 22–25 days operational and 50%–60% capacity utilization, while the net earning is around US$200–$250 per month (Saran et al. 2013).

7.2.2.2 Fruit Leather Preparation

Studies to standardize the technique for preparing fruit leather by blending pulp from two indig-enous fruits–i.e., Bael (*Aegle marmelos*) and Aonla (*Phyllanthus emblica*)—were funded by the Leverhulme Trust, UK.

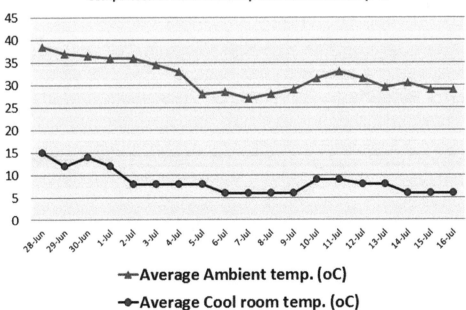

FIGURE 7.2 Comparison of ambient and cool room temperature (°C) at Sultanpur.

Aonla and bael, two highly nutritious fruits with a very high medicinal and therapeutic value, are normally not used as table fruits. Making a leather preparation of blended bael and aonla is a low cost technique because, being underutilized, these fruits are available in the market at a low price. Besides, the leather is easy to handle and distribute, and requires no specialized storage. Due to its appealing taste, it is immensely acceptable to children and therefore can be included as a nutritious dessert in the midday meal (Dwivedi et al. 2015). In this project, fruit leather packed and stored in laminated pouches showed better storage quality in terms of nutrition and acceptability during 90-day storage when compared to polythene pouches. The fruit pulps were blended using nine combinations of bael, aonla, and sugar. The best ratio for bael and aonla leather was 2:1 bael to aonla with 10% sugar and 2000 ppm Potassium metabisulphite (KMS), in respect of shelf life and quality (Dwivedi et al. 2015).

7.2.3 EXTENSION INITIATIVES

7.2.3.1 Farmer Training and Demonstration at Three Districts of Uttar Pradesh

"Rural bio-resource innovation-application to uplift the socio-economic status of farmers and entrepreneurs from Uttar Pradesh" was a project funded by the Department of Biotechnology (DBT), Government of India.

The project objectives were:

1. Address the postharvest technology issues pertaining to loss reduction, proper marketing of the fresh produce after sorting, grading, and packing, developing storage and transportation facilities, primary processing, minimal processing and semi-processing to reduce spoilage, utilization of waste generated during handling and processing, and providing postharvest management training and demonstration to farmers and entrepreneurs, with emphasis on rural women
2. Popularize and disseminate bio-control integrated pest management (IPM) strategy through awareness and training programs to the targeted population through field study and demonstration
3. Demonstrate animal husbandry and extension services with critical inputs to the livestock farmers, and introduce nutritious grass spp. (viz. Hybrid Napier, Deena Nath grass (*Pennisetum pedicellatum*), Anjan grass (*Cenchrus ciliaris*), Guinea grass (*Panicum maximum*), Dhawalu grass (*Chrysopogon fulvus*), and *Azolla* as sources of green animal feed

The work started with a preliminary survey and site selection. The sites selected were (1) Gayatri Suman Farm and Nursery, Behta, Bulandshahr, (2) Krishi Vigyan Kendra (KVK), Pursi, Murad Nagar, Ghaziabad and, (3) Krishi Vigyan Kendra, Chhollas, Dadri, Gautam Budh Nagar.

At each of the selected sites, a Common Facility Centre (CFC) with the following facilities was established: Processing equipment such as: (1) fruit and vegetable washing tank, (2) baby pulper, (3) fruit crusher, (4) basket press for juice extraction, (5) crown corking machine, (6) mechanical dryer, (7) screw type juice extractor, (8) multipurpose vegetable cutter, (9) pouch sealing machine, (10) gas stove and cylinder, (11) weighing balance (3 and 30 kg capacity), (12) stainless steel sieve and plastic crates (25 of each), (13) refractometer, (14) bottle washing machine, (15) gas burner, (16) solar power fan with inverter, (17) thermometer (Zeal brand), and (18) stainless steel vessels of different sizes, as well as other necessary utensils for processing were also provided to the three CFCs. Besides equipping the CFCs with processing laboratories, some ancillary facilities were also provided (viz: ZECC of 100 kg and 1 MT capacities, vermicompost, NADEP, and *Azolla* pits).

A total number of 1508 farmers and entrepreneurs were provided hands-on training and demonstration at the three fully equipped CFCs. The targeted areas for training were:

1. Postharvest management of fruits and vegetables
2. Biocontrol of crop diseases and pests, including root-knot nematode
3. Planting of nutritious grass spp. and cultivation of *Azolla pinnata* as green animal feed

In postharvest management of fruits and vegetables, the main focus was on, (1) storage of locally grown raw fruits and vegetables in ZECC of 100 kg and 1 MT capacities to extend shelf life and reduce losses, and (2) conversion of raw agricultural produce into value-added products (viz. purees, juices, squashes, nectars, and juice blends). Food safety, hygiene, and marketing requirements were also conveyed to the farmers and entrepreneurs.

For biocontrol of diseases and pests, and infestation of root-knot nematodes in crops, applications of biocontrol fungi (viz. *Trichoderma harzianum*, *Beauveria bassiana*, and *Paeciliomyces lilacinum*) were demonstrated. Farmers were provided cultures of *Trichoderma* and trained how to prepare soil amendments.

In the animal husbandry program, farmers were given seed packets of five forage grasses (viz. Hybrid Napier, Deena Nath grass, Anjan grass, Guinea grass, and Dhawalu grass). They were asked to grow them on fallow land and use them as cut grasses to feed the cattle. Farmers were also shown how to make pits for the cultivation of *A. pinnata* for use as animal feed. Starter cultures were provided to farmers for cultivating *Azolla*.

7.2.3.2 Postharvest Technology Training at KVK-Sylvan, Hengbung Senapati District, Manipur, North East India

The DBT project was extended to Manipur, a state in the northeast in which seven hands-on trainings were conducted at KVK-Sylvan, Senapati district. The total numbers of participants were 237 tribal youth, out of which 217 were females and 20 were males. Each training program lasted three days. The training was focused on postharvest handling and processing of fruits and vegetables with an emphasis on food safety.

Some of the processed products included primarily processed pineapple (*Ananas comosus*) and cauliflower; semi-processed plum pulp (*Prunus domestica*), pineapple pulp, and passion fruit juice (*Passiflora edulis*); and fully processed products (viz., ketchups, toffees, pickles, juices, squashes, nectars, etc.) using various fruits like banana, Indian plum, pineapple, mango, passionfruit, tomato, potato, and mixed vegetables.

The participants were very enthusiastic about the training program. Among the trained participants, two became entrepreneurs and some of the products were used in the development of a family enterprise.

7.2.4 EDUCATION INITIATIVES

7.2.4.1 E-Learning Project with the Postharvest Education Foundation, United States

7.2.4.1.1 South Asia Postharvest E-Learning Program

Amity University, in collaboration with the Postharvest Education Foundation (PEF), launched a pro bono one-year d-learning program on postharvest management for South Asia, in which 30 candidates were selected after a screening of applications. The course material was prepared by PEF and the program was run from e-learning Centre, Amity University. The program included reading assignments, website resources, PHL assessments, quality evaluation, commodity systems assessment, small-scale postharvest handling practices, cost/benefit analysis, and a variety of homework and fieldwork assignments to be conducted in the participant's local area.

As the concluding part of the program, a 3-day workshop was organized, August 6–8, 2013, at AICPHT and CCM, in which 14 successful candidates—6 from India, 1 from Iran, and 7 from different South Asian countries (viz. Pakistan, Sri Lanka, Bangladesh, and Nepal)—were invited for hands-on training in some important areas of postharvest management of perishable fruits and vegetables. Besides the 14 participants from the e-learning program, 14 additional participants were registered from India, giving preference to industry, academia, Farmers' Self-Help Groups (SHGs) and small and medium-size enterprises (SMEs). A total of 28 participants took part in the Training of Trainers (ToT) program in the specific area of integrated postharvest management of fruits and vegetables, targeting reduction in PHL. The three-day workshop covered the following areas through laboratory training and lectures by specialists:

- Proper marketing of fresh produce after sorting, grading, and packing
- Developing storage and transportation facilities
- Improved container with ventilated liners
- Use of shade
- Field packing systems
- Low energy cool storage: ZECC
- Low cost, CoolBot equipped small rooms
- Field curing of root and tuber crops
- Improved solar drying methods
- Low cost food processing
- Utilization of leaves of vegetable crops for obtaining β-Carotene (proitamin A) and other micronutrients and minerals in powder form, to be used as dietary supplements and as animal feed
- Primary processing, minimal processing, and semi-processing, and appropriate packaging to fetch a better market price
- Modified Atmosphere Packaging (MAP) of fruits and vegetables

7.2.4.1.2   Impact of the Program: Outcomes

- Testing the applicability of the e-learning model in Integrated Postharvest Management (IPHM). The sub-objective here is the testing, at a smaller scale, of the integrative functionality of the learning platform, and to probe its potential for generalizations
- Design, development, and implementation of e-learning system for the farm trainers, giving emphasis to the greater spread and use of IPHM technologies in everyday life
- Development, through the integrated system, of technical and human capacity to implement the overall process of knowledge management in the organizational context of the rural sector
- Creation of a farmer training network by implementing the modern practices to increase market share, reduce costs of production, and maximize utilization of the produce (up to 100%)

7.2.4.2   Postgraduate Degree Program

7.2.4.2.1   MSc Horticulture (Postharvest Technology)

The postgraduate program was started in 2014–2015 and has been designed in coordination with both national and international experts. The degree program combines classroom teaching with strong practical work, field, market, and industrial visits, as well as interactive sessions with eminently successful persons in respective areas to prepare students/trainees to efficiently tackle problems related to horticultural losses as scientists, technocrats, trainers, consultants, and entrepreneurs. The research of these students is focused on different value-added products from fruits, vegetables, and mushrooms, development and standardization of storage and packaging technologies, and edible coatings to enhance the shelf life of different horticulture produce. Two batches of students have already passed and are working in different government and private sectors.

7.2.4.2.2   PhD Horticulture with Specialization in Postharvest Technology

The center has six PhD students who are working on different aspects of storage, processing, postharvest quality evaluation of conventionally and organically grown vegetables, and packaging technologies with crops like strawberry, bitter gourd, tomato, lilium, etc.

7.3 CONCLUSIONS

Although having ample production, the supply chain mechanism has been unable to cope with the flood of farm produce. This inefficiency in the supply systems is a constraint, which in turn has a direct impact on inflationary pressures and degradation in the produces' nutritional quality. In effect, production alone is not sufficient to ensure the reach of food to our dispersed subcontinental footprint. The missing piece for achieving food security is a good distribution mechanism.

Greater value to farmers will arrive by assigning emphasis on post-production activities that connect the farm harvest to markets for value realization. This will include expanding the marketing range of the farmers. Reducing losses in the postharvest supply chain and providing pan-India marketing are the most important steps.

The key strategies behind policy interventions that aid post-production market linkages are:

- Promote direct access by farmers to all avenues to monetize their produce and expand their market outreach
- Organize postharvest collection activities at the farm gate so as to build capacity to minimize the handling loss and convert would-be-loss into value
- Promote private sector participation in expanding the reach and range of farm produce into consumption center, both domestic and international

Although the factors affecting PHL are well understood and technologies to combat them have been developed, they have not been implemented, in many cases, due to one or more of the following reasons:

- Inadequate handling and marketing systems
- Inadequate transportation systems
- Unavailability of tools and/or equipment
- Lack of appropriate information
- Inadequate cold storage and reefer vans
- On-farm training mechanisms

An integrated approach for postharvest science and education from high school through the post-graduate level could help to reduce global food losses by integrating postharvest information into the general agricultural curriculum of each country or state, strengthening the extension services, with major emphasis on preventing losses, maintaining quality and nutritional value after harvest, and ensuring food safety. Capacity development efforts for postharvest technology in developing countries must be more inclusive, including technical knowledge on handling practices, access to postharvest tools and supplies, cost/benefit ratios, augmenting the knowledge of all stakeholders, skill development (training needs assessment, teaching methods, and advocacy), and establishment of collection centers and packhouses at strategic locations.

REFERENCES

Dubey, N. and Raman, N. M. (2016). Low-cost insulation for coolrooms using 'CoolBot'. *Acta Horticulturae*, 1120, 279–284.

Dubey, N., Roy, S. K., Saran, S. and Raman, N. M. (2013). Evaluation of efficacy of zero energy cool chamber on storage of banana (*Musa paradisiaca*) and tomato (*Solanum lycopersicum*) in peak summer season. *Current Horticulture*, 1(2), 27–31.

Dwivedi, S. K., Mishra, V., Saran, S. and Roy, S. K. (2015). Studies on preparation and preservation of fruit leather by blending indigenous fruits viz bael and aonla pulp. *Journal of Postharvest Technology*, 3(2), 36–42.

Indian Horticulture Database (2016). National horticulture board. Ministry of Agriculture, Government of India. Retrieved from www.nhb.gov.in

Kitinoja, L., Al-Hassan, H. A., Saran, S. and Roy, S. K. (2012). Identification of appropriate postharvest technologies for improving market access and incomes for small horticultural farmers in Sub-Saharan Africa and South Asia. Invited paper for the IHC Postharvest Symposium, Lisbon, August 23, 2010. *Acta Horticulturae*, 934, 31–53.

Roy, S. K. (1985). *Zero Energy Cool Chamber. 1985. Research Bulletin No. 43*. New Delhi, India: Indian Agricultural Research Institute.

Saran, S., Dubey, N., Mishra, V., Dwivedi, S. K. and Raman, N. M. (2013). Evaluation of shelf life and different parameters under CoolBot cool room storage condition. *Progressive Horticulture*, 45(1), 115–121.

8 Training Women on Reducing Postharvest Losses of Fresh Fruits and Vegetables in Egypt

Saneya Mohamed Ali El-Neshawy

CONTENTS

8.1 THE PRESENT SITUATION OF WOMEN'S INVOLVEMENT IN POSTHARVEST ACTIVITIES AND THEIR KNOWLEDGE GAPS

Women's roles have traditionally been confined to carrying out household chores, including raising children and caring for elderly family members. Their involvement in agriculture production activities was limited to household farms, which included seed sowing, cattle rearing and milk production, and the sale of poultry products, or working in produce packing and in small enterprises. The data available on women's contributions in postharvest activities are unreliable and outdated. There is a need to systematically identify the information needs of the rural women in various communities, as their needs are varied. Very few studies have assessed the information needs of women towards the activities related to postharvest management of fresh fruits and vegetables.

Women's contributions to postharvest management and reducing losses are still the key factor to increasing the demand for produce, which in turn would elevate the family income. In order to get better acquainted with the relevant information, they are keen to attend training sessions on agriculture practices in general, and on postharvest operations and systems in particular (El-Neshawy and El-Sayed 2005).

Through primary sessions to evaluate their ability to join training programs, it was found that the performance of women in agriculture practices and as postharvest workers is constrained in many ways. They have less access to information in comparison with men because they are not encouraged, or even invited, by agricultural department staff to participate in extension programs. Women usually collect some information from radio talks, television programs, and from neighboring private firm workers.

While exploring and acting upon the question "How would their situation change if rural women acquired the knowledge to share in postharvest operations throughout the postharvest handling system?" it became evident that, for the women to gain knowledge, as it would mean changing their status.

8.2   NEED FOR THE TRAINING OF WOMEN

The women selected as participants in the training sessions were individuals who were involved in postharvest research careers or women working as partners in different postharvest operations, with a certain level of education that qualified them to be ready to gain knowledge and get acquainted with the latest technologies.

The participant women's needs were assessed through in-depth interviews, which indicated that the lack of information on acquiring suitable postharvest technology and modern operations was the primary missing element. Needs assessment surveys were conducted using questionnaires to determine the exact information needs of women in agriculture, and in postharvest technology in particular.

During interviews, women were asked to reflect on their aspirations and to identify what information and knowledge they needed (Buskens 2014). These reflections highlighted a large gap between the present status and their experiences in the postharvest field, which basically contributed to the existing postharvest losses (PHL). The training programs and interactions among the women during the training could affect the women in many dimensions, helping them to become successful workers in the postharvest handling system, and in export companies they may join.

8.3   TRAINING PROGRAM

The ultimate target of the training program is to make information available to rural women to empower them to be more oriented towards the use of improved postharvest technology for reducing PHL. Women's access to training programs is very limited in rural areas, since the training activities are conducted at distant city centers. For the women to have proper access, the training should be conducted in the nearby villages, in the local language they prefer, and in accordance with a schedule convenient to them. The training sessions conducted for the women were well-appreciated for the use of posters, videos or computer presentations, workshops, and group discussions to identify, review, and prepare the needed information. The programs have constantly evolved through the cycles of learning, implementation, sharing, and reflecting.

Determination of the preferable methods of training is essential to define the training needs of rural Egyptian women in implementing certain agricultural activities. It is best advised for woman scientists to provide training for young women in Egyptian villages on topics like harvesting, sorting/grading, packaging, storage and marketing of horticultural crops. On-field demonstrations of harvesting techniques, postharvest technologies, and packhouse visits would assist in proper understanding to allow women to apply the extracted technologies in their own fields. Women should also be instructed in the field at the latest stages before harvest, for example, on how to judge proper ripening before the harvest. Extension and consultancy services are conducted regularly to identify and solve problems faced during the crop production cycles.

FIGURE 8.1   Explaining harvest quality of mango fruit to the trainees.

It has been observed that the women who receive such training in the village are often better motivated and further willing to take other training courses at the centers or at the packinghouses.

Throughout my career as a postharvest specialist, I offered support to rural women through extension work in fields, in packinghouses, or through companies to give instruction and advice. The women were trained in good postharvest practices, application of the ideal methods of quality indices for harvest, and quality maintenance throughout the supply chain, to allow the fresh produce to be delivered to consumers and to increase the scope for the export (Figure 8.1).

The effects of training programs can be observed through:

- The increased production of high-quality agricultural produce by implementing the knowledge acquired and practicing the new postharvest technology to reduce PHL. These will, in turn, boost the net income of the producers
- An enhanced relationship between the rural farmers and cooperative associations at the market yards, with updated market information

8.4   ORIENTATION LECTURES TO THE WOMEN EXPLAINING THE NECESSITY OF POSTHARVEST TECHNOLOGY

The training programs created a good opportunity for rural women to connect with each other, to identify and share problems, and to discuss solutions for different topics, including postharvest aspects, modern technologies, and the ways they convinced their neighbors to follow (Figure 8.2). The women trainees were regularly tested for their knowledge of postharvest handling, pathology, and quality aspects using a series of questions about the best practices at different levels of the supply chain. The training showed positive results, and all the trainees were increasing their awareness of the proper steps and operations throughout the postharvest handling system that served to maintain quality and minimize food losses.

FIGURE 8.2   Training women on postharvest aspects.

8.5   DESIGNING TRAINING PROGRAMS SUITABLE FOR WOMEN WITH DIFFERENT BACKGROUNDS

The training programs were designed after proper identification of what would enable the women to increase crop productivity through gaining knowledge of new postharvest technologies, which guaranteed the best quality and reduced PHL.

Based on the in-depth interviews, the trainer's team designed an extension program named: "Enlightening rural women of high-quality fresh fruits and vegetables with reduced postharvest losses."

The focus of the training program was on the strategies to produce and maintain high-quality fresh produce suitable for export, starting from the harvest and continuing until it reached the end users (from "field to fork"). The topics included crop production, pre/postharvest aspects including postharvest handling system, marketing, funding availability, and the agricultural cooperative association.

8.5.1   Postharvest Food Losses

Losses from postharvest spoilage are estimated to be over 50% in developing countries, including Egypt. In brief, we don't need to clear large areas for farming, while discarding produce, which eventually results in food losses and wastes. By preventing food spoilage and reducing PHL collectively through the private sector, non-governmental organizations (NGOs), and the government, we can protect the environment. Establishing training centers with the necessary agro-handling facilities within the farming communities would make available qualified personnel in the related fields for the farmers' service, while also charging the farmers an affordable and reasonable fee. A lot has been taught to women farmers on the major elements that cause food losses and waste of fresh fruits and vegetables. In that way, farmers would look at agriculture as a business, not as a subsistence activity, and more people will be encouraged to get trained in food production and proper postharvest handling to reduce PHL.

8.5.2   Monitoring the Important Factors Causing Food Losses and Waste after Harvest (Examples)

Improper packaging could cause serious PHL in fruits and vegetables:

- Packages that could cause wounds and cracks in citrus may further lead to green mold, caused by *Penicillium digitatum*.
- Strawberries packed in wooden boxes experience wounding and bruising.

- Displays of fresh citrus fruits "on the street" for marketing are exposed to the sun, and are thus subject to desiccation and quick deterioration.
- The packaging of potatoes in rough fabric materials expose them to contamination or wounding.
- Overloaded trucks cause impact wounds and lead to a loss of bananas.

8.5.3 Practical Assessment of Food Losses and Waste in Fresh and Dry Produce

The assessment of food losses and waste in fresh and dry food is not recorded, as for the total fresh or dry production on the country level; this means that no numbers for the total losses or waste percentage were precisely assessed. Studies are underway for assessment of losses to be accounted individually, but not gathered for the whole postharvest chain. It is estimated in general to be about 50% of the total production, including the average of the waste on each operation or step, or as the rejected quantity of the exported lots.

8.5.4 Methods to Reduce Food Losses and Waste

Fungicides have been a major weapon in combating PHL; however, because of health and environmental concerns those synthetic fungicides pose the greatest carcinogenic risk among pesticides on food. Most of the major fungicides used for postharvest diseases control have been taken off the market.

Control methods of postharvest diseases of fruits and vegetables using safe alternatives to synthetic pesticide use are the most important factor for reducing food losses. The risk to consumers, as well as the entire ecosystem, from the use of synthetic fungicides in recent years, and the need to find alternative methods of control of plant pathogens, including biological control (Wilson and Chalutz 1989), is well taught to rural women.

The use of naturally occurring antagonistic microorganisms is a potentially new form of fruit protection as an alternative to chemical use. It considered one of the most important, effective, and promising methods of using biological control as a control strategy. Several biocontrol agents have been developed and widely investigated against different postharvest fungal pathogens (*Penicillium* spp., *Monilia* spp., *Alternaria* spp., *Botrytis cinerea*, and *Rhizopus stolonifer*), and tested on various commodities such as citrus, grape, strawberry, apple, or pear (Wilson et al. 1993; Wilson and Wisnieski 1994).

8.5.5 Precautions and Personal Protection of Women Workers and Technicians in Horticulture Production and the Supply Chain

Women were instructed on the mistakes that may occur with the products prior to packaging, such as leaving the decayed ends of sweet potatoes, which would serve as a source of infection to other tubers during shipping. They were also instructed to sanitize the wash water before hand washing the products during preparing for palletization.

8.6 WOMEN'S AWARENESS

The awareness of rural women on the importance of postharvest technologies in the production chain in developing countries like Egypt is important, as the majority of producers focus only on the quantity produced, but not on the quality. To increase women's awareness of the necessity of quality management to be followed during their work in the postharvest handling system, training programs were conducted on the following topics:

- Awareness toward orchard sanitation and the risk of burying infected or deteriorated fresh fruits in the soil near the farm, as the infection can be transported back to the farm as spores via irrigated water. The risk of infection is particularly high in vegetables and fruits in contact with the soil
- Commitment to proper and healthy procedures of postharvest handling systems, starting from picking and continuing to the end users

- Maintaining an adequate cool chain, along with the whole handling processes and supply chain (Quality in Chain [QUIC])
- Commitment to hygiene procedures, such as wearing masks, gloves, hat cover, etc., to guarantee the safety of products and human workers
- Harvest methods of fresh fruits and vegetables, in terms of the best way of harvesting
- The disasters of pesticide use and the importance of effective safe control methods when used as pre- or postharvest treatments at certain concentrations, and with regard to its dose and residual effects on human health. This topic included focus on fungicidal residues in fruit tissues, in particular those with a high water content that are harvested fresh and consumed immediately after harvest
- Food contamination with mycotoxins and or chemical fungicide residues that last in fresh fruits and vegetables. Instruction for home economic and public health aspects to household women are essential
- Fungal deterioration in stored seeds and grains, aflatoxin production, and the risks of its development regarding the harmful effects on humans, including explaining the methods of detoxification during storage
- Biological and physical methods in controlling postharvest diseases, including the use of biological agents such as yeasts or bacteria
- Influence of Modified Atmosphere Packaging (MAP) through gas modification, including reducing oxygen and increasing carbon dioxide levels in the storage atmosphere to reduce respiration
- Influence of MAP on aflatoxin production in grains, and suppression of toxin production as a result of fungal suppression
- Obtaining and maintaining high-quality products through applying the new technologies of postharvest handling chain to reduce PHL, which will reflect positively on the net income of the producers
- Acquiring knowledge and obtaining information required in the field of postharvest diseases and related aspects, including traceability of exported fresh produce

8.7 WOMEN AND INFORMATION AND COMMUNICATION TECHNOLOGIES (ICTs) FOR EMPOWERMENT

The role of ICTs in rural women's empowerment is based on providing information required by them through websites designed by the organizations conducting the training programs. The experience of using the ICTs expanded the women's knowledge and helped them in applying modern techniques in the postharvest field to guarantee good quality of the produce, reduce food losses, increase export opportunities, and improve social standards.

Most of the women in rural areas have very little access to information, as they are usually illiterate, and too economically weak to afford even the basic forms of ICTs, such as radios and telephones. Hence the ICTs are expected to be affordable and user-friendly, to enable the women to use them to get accurate information efficiently. However, gender consideration has not received the required attention while designing the information services or application of ICTs.

The potential of ICTs to meet the information needs of women will no doubt offer several channels for the exchange of information and numerous opportunities for rural women farmers.

Moreover, current mass media and communication systems have not been used to their maximum capacity for social development. Information should be accessible to rural women at selected sites, with various ICTs to facilitate easy access to relevant information and information exchange.

The lack of reliable and comprehensive information for women is a major hindrance to agricultural development. They require information on agricultural inputs, market prices, transportation systems, product potential, new environmentally sound production techniques and practices, new agricultural technologies, new markets, food processing and preservation, decision-making

FIGURE 8.3   Mentoring trainees on computer skills and to browse internet for agriculture information.

processes, the resource base, trade laws, and trends in food production, demand, and processing. Women also need to exchange indigenous knowledge. However, most available local information is packaged in a raw form and is therefore difficult to access or use (Paquot and Berque 1996). The situation is compounded because women do not know where to find this information.

The lack of information relating directly and indirectly to farming strategic crops and becoming farm managers was addressed through a web-based program via Gender Research in Arab Countries into information Communication Technology for Empowerment-Middle East and North Africa (GRACE-MENA), which was also made available on CD, as a paper document, and through my provision of training, including regular visits to the women's fields. The web program was found useful in many ways.

Rural women were mentored on computer skills, ways to search, and how to use the web-based program's content, as well as other online postharvest information and knowledge resources. Women were provided with a CD-ROM version of the Internet-based information to use when their Internet connection was limited or not available. Regular refreshment training was conducted for rural women for better learning, implementation, and sharing of the information related to maintaining and preserving fresh produce from the spoilage and deterioration that cause huge losses (Figure 8.3).

Quizzes and tests with yes or no, right or wrong, and or give a reason(s) answer options were conducted, and scores were given at the end of the training courses to encourage those who will teach others and for the consolidation and sharing of information.

8.8   CASE STUDIES

Previously, a study was conducted for women farmers in order to determine whether and how the use of ICTs, in the form of a web-based agricultural extension program, could support the women's objective of taking on the management of their own land (Sharma 2001; Bakesha et al. 2009). The study concluded that web-based agricultural extension programs enhanced women's agricultural knowledge and marketing opportunities.

A long-term training program was provided in Egypt on the use of ICTs to assist female landowners to become active farmers on their land. The lack of information that related directly or indirectly to farming strategic crops and becoming farm managers was addressed by offering a web-based training program. Based on in-depth interviews for women, a need-based extension program with content that included detailed information on crop production, marketing, funding availability, land rights, landowner credit cards, registration of land credit, and the agricultural cooperative association was designed. A team of scientists and specialists worked as trainers to mentor women on computer skills and on how to search and use online agricultural information resources and other web-based content. The program evolved through cycles of learning, implementation, sharing, and reflecting.

As the women farmers become confident in their capabilities, and with the increased knowledge of agriculture production and postharvest procedures, their incomes were markedly increased. The training created an opportunity for the women farmers to connect with others, which helped them to be successful with the power of decision-making over their small enterprise. Such networking allowed the women to recognize issues related to postharvest technology and their solutions. With the change in women's behaviour, agricultural productivity increased, and family well-being also improved (El-Neshawy 2014).

8.9   CONCLUSIONS

With these successful live and ICT-based training programs, women in Egypt are becoming well-trained on the detailed steps of improved postharvest technology. They have been able to reduce losses due to poor handling and to produce high-quality products to meet the export requirements of foreign markets. Moreover, with this appropriate training in technical competencies or simple management practices, women who wished to have a postharvest job or create a microenterprise were able to market their produce through existing cooperative associations.

REFERENCES

Bakesha, S., Nakafeero, A., and Okello, D. (2009). ICTs as agents of change: A case of grassroots women entrepreneurs in Uganda, In: I. Buskens and A. Webb (Eds.) *African Women & ICTs Investigating Technology, Gender and Empowerment*, pp. 143–153, Zed Books, London, UK.

Buskens, I. (2014). A methodology of love. In: I. Buskens and A. Webb (Eds.) *Changing Selves, Changing Societies: Women, Gender and ICT in Africa and the Middle East*, Zed Books, London, UK.

El-Neshawy, S. (2014). Transforming relation and co-creating new realities: Landownership, gender and ICT in Egypt. In: I. Buskens and A. Webb (Eds.) *Women and ICT in Africa and the Middle East: Changing Selves, Changing Societies*, pp. 275–287. Zed Books, London, UK.

El-Neshawy, M.S. and El-Sayed, A.A. (2005). Women participation in production, marketing, and storage of fresh horticulture crops for local consumption and export, International conference on women's impact on science and technology in the new millennium, Institute of Science, Bangalore, India.

Paquot, E. and Berque, P. (1996). Summary of the achievements of the CTA since the seminar Montpellier, 1984. In: *The Role of Information for Rural Development in ACP Countries: Review and Perspectives: Proceedings of an International Seminar, June 12–16, 1995, Montpellier, France*. Technical Center for Agriculture and Rural Cooperation, Wageningen, the Netherlands. pp. 61–83.

Sharma, C. (2001). Using ICTs to create opportunities for marginalized women and men; the private sector and community working together, Paper presented at the World Bank, Washington, DC, December 18.

Wilson, C.L. and Chalutz, E. (1989). Postharvest biology control of Penicillium rots of Citrus with antagonistic yeasts and bacteria. *Scientia Horticulture*, 40, 105–112.

Wilson, C.L. and Wisniewski, M.E. (1994). *Biological Control of Postharvest Diseases of Fruits and Vegetables -Theory and Practice*. CRC Press, Boca Raton, FL, p. 182.

Wilson, C.L., Wisnieswski, M.E., Droiby, S., and Chalutz, E. (1993). A selection strategy for microbial antagonists to control postharvest diseases of fruits and vegetables. *Scientia Horticulture*, 53, 183–189.

9 Investing in Postharvest and Food Processing Training, Equipment, and Value Addition for Reducing Food Losses in Tanzania

Bertha Mjawa

CONTENTS

9.1 INTRODUCTION

9.1.1 MAGNITUDE OF POSTHARVEST FOOD LOSSES IN TANZANIA

Food crop losses in Tanzania have been reducing farmers' incomes, thereby increasing poverty. For example, the postharvest losses (PHL) in the horticultural sector (especially fruits) could be as high as 60%, due to lack of proper collection, storage systems, processing, packing, and preservation facilities. Losses in cereal crops, legumes, and nuts are reported to reach 30%–40%, due to malpractices (viz., late harvesting, rudimentary methods of handling, poor storage facilities, inefficient tools for transportation, processing, and packaging; [Abass et al. 2014; Abdoulaye et al. 2016]). PHL in roots and tubers reach 45% of total produce, due to poor harvesting practices, packing, transportation, and handling tools that cause bruising and rotting. Losses in the dairy and fisheries sectors reach as high as 80%, due to improper handling (Diei-Ouadi and Mgawe 2011; FAO 2011).

Postharvest handling and management trainings to empower smallholder farmers would improve the quality of horticultural crops, cereal grains, oilseeds and nuts, pulses, and animal products, and in turn assist to improve their profits by selling higher quality produce.

Investments made at different levels of the value chains, in the form of education, equipment, and training covering subsystems of harvesting, threshing, sorting, grading, transportation, storage, processing, appropriate packaging, and marketing will lead to reduction of physical and economic PHL in many value chains.

Government interventions following enormous losses in postharvest systems set various priorities (e.g., achieving food security), strategies, programs, and projects with a focus on public–private–producer partnerships (4Ps) to reduce these losses (Barreiro-Hurle 2012; SAGCOT 2010; URT 2016a and b; MIVARF 2017).

Some of the major programs and projects include Agricultural Sector Development; Marketing Infrastructure Value Addition and Rural Finance Support; a few projects dealing with post-production, like United States Agency for International Development (USAID) Feed the Future, national projects like Muungano wa Vikundi vya Wakulima Tanzania (MVIWATA) and Agricultural Non-State Actors Forum (ANSAF), and non-governmental organizations (NGOs).

9.2   A CASE STUDY OF MARKETING INFRASTRUCTURE VALUE ADDITION AND RURAL FINANCE SUPPORT (MIVARF) CAPACITY BUILDING ACTIVITIES IN TANZANIA

A seven-year program has provided capacity building to a total of 37,500 beneficiaries from the 37 districts of Tanzania. Key areas of engagement included improved food processing machinery through matching the grant, rehabilitation and equipping postharvest training centers, and training in postharvest management as well as food processing strategies. Other areas that synergised the value addition for these beneficiaries include producer empowerment of smallholder groups, which led to the availability of raw materials to these processors, and enhanced access to financial services for enabling producers to engage in a profitable business. A key objective was to ensure that farming is practiced as a business and not as subsistence for the farm families. This book chapter provides a case study of one group of beneficiaries that invested in food processing, and whose investment led to improving lives of farm families as well as reduction of food losses.

9.2.1   Project: The Grande Demam Dairy Processing Industry

Grande Demam Dairy processing factory started its activities in 2013. The dairy processing factory is located in Usa River, Arusha. The project was an outcome of a concern by Dr. Deo Temba (founder), a veterinary scientist who was a dairy farmer and veterinary doctor. Dr. Deo kept four dairy animals, which produced about 40 L of milk per day, while his daily home consumption was about five L. Moreover, most of his neighbors also kept at least one dairy cow, thus the expectation of selling his surplus milk to the neighbors was not possible. As a veterinary scientist, Dr. Deo interacted with many dairy farmers in the region who also shared the same experience of having surplus milk. It was from that experience that Dr. Deo thought of supporting farmers by installing a cooling facility with a capacity of 200 L to store the milk collected while looking for a market in Arusha city. Overwhelmed by the increasing volume of milk at the cooling facility, in August 2012, together with support from his staff, the facility started basic processing of cultured milk, also commonly known in Kiswahili as *mtindi*. The market was assured, as the processed milk had a longer shelf life. They were therefore encouraged to expand operations.

Encouraged by the sale pattern of the *mtindi*, the concept was quickly translated into the birth of Grande Demam Project in January 2013. After spending approximately three months developing the project systems, including looking for a market for the *mtindi* milk, the project realized it could not penetrate the market and needed a more innovative product to gain a wider market. At the same time, the situation at the cooling plant worsened, with a larger supply of raw milk as compared to the processed milk delivered in the market.

Working with about 76 small-scale dairy farmers, the project quickly consolidated its operation to about 180 L of milk per day. By the end of 2013, the farmers attached to the project reached 100 and produced an average of 350 L per day. In the year 2014, the project purchased additional machines and equipment, which enabled it to expand its operation to 600 L per day and introduced contract arrangements. This enabled the farmers to deliver milk to the project and receive payments at two-week intervals. As part of the contract arrangement, the farmers also got access to flexible veterinary services, including drugs and supplements at affordable prices, payable after receiving their revenue from the supply of milk. By mid-2015 the project expanded its services to Meru and Siha districts, and increased farmer outreach to 200, producing about 800 L of milk per day.

The vision of the Grande Demam Dairy Project seeks to be the leading dairy processing facility in northern Tanzania. This vision is supported by three main pillars (i.e., a commitment to increase farmer support services, improved milk collection and processing, and effective marketing). The mission of the project is to develop the capacity of the small-scale dairy farmers to adopt and develop the dairy industry to levels at which it can meet the needs of the people on a sustainable basis.

The ultimate goal is to enhance the capacity of the farmers to improve their livelihood security, health, and poverty alleviation, using dairy technology. Both goals, the vision, and the mission lead to PHL reduction of milk produced by about 4000 farm families (increased from 800 in 2016) of Meru District, Arusha Region through scaling up the dairy business outreach activities and through innovations driven by the economic necessity to accelerate trading, investment, and sustainability. The whole philosophy of this small-scale processing industry is to become a competitive multi-dairy business investment from its initial idea of transformation and diversification. A shift in operational processes has now been made possible by working with strategic partners who share the dream of economic prosperity of the small-scale dairy farmers in Arusha region. The important partner, in this case, has been MIVARF, who supported the small-scale industry in the area of matching grants, training, farm visits, exhibitions, and outreach activities.

The assumption was, and still is, that processing of dairy products has strong appeal in the industry, and is the engine of competitiveness and fair allocation of resources for a farmer-led approach to scaling up dairy farming activities in the region. Consequently, the desire of Grande Demam project is to drive transformation in the dairy sector (e.g., increase milk productivity, lower unit dairy processing costs, and above all increase farmers' incomes and business sustainability).

Some of the activities implemented to achieve the abovementioned goal include:

- Expanding dairy production to accelerate the poverty alleviation process among 2,500 small-scale dairy farmers in Meru area
- Improving the capacity of the project to step up animal health control programs among dairy farmers
- Improving dairy processing by the industry, using modern state-of-the-art equipment

9.2.2 Partnership with MIVARF

Overall, the partnership with MIVARF has enhanced competitiveness in the dairy industry value chain. Utilising this opportunity effectively, the Grande Demam developed a project proposal to catalyze evolution and be able to effectively engage in modernized dairy farming and trade relationships. The desire was driven by three interrelated objectives: firstly, to reorganize milk production and market supply on the side of the producers, using modern milk collection and transportation technologies; secondly, to eliminate marketing bottlenecks, including inadequate financial capacity, which impede the operations, transaction costs, scope, and quantity of milk collection and dairy product market penetration; and thirdly, to develop effective management of the project, including trade relationships, to enable the project to engage as leaders in the supply of high-quality products in the modern dynamic dairy market. However, despite its determination and alertness, Grande Demam took cognizance of the fact that the dairy farming scaling-up agenda was faced with critical issues of supply-side inadequacies in technology, capital investment, resource base, and marketing facilities.

The desire to improve these dairy activities has brought with it the need for negotiating for strategic capacity from MIVARF to enable the project to support the farmers with better dairy technologies and to have the capacity to respond to the highlighted gaps. It is for this reason that the Grande Demam processing industry declared intent to partner up with MIVARF to stimulate the agenda bubbled as "Economic Prosperity using Dairy Technology."

MIVARF has a matching grant window that provides rural business processors with a 25%:75% funding to procure efficient machines. The mode of engagement with these processors is that the beneficiaries are in business, have a market for their products, but lack technology both in terms of machinery and knowledge. Rural processors use this window, coupled with MIVARF capacity building opportunities, to acquire efficient machines as well as knowledge in postharvest management and processing.

Grande Demam applied for this fund in November 2016 and has undergone all the procedures, including a business plan, a qualified premise that is certified by the national food and drug authority, and an application form. Following these initiatives, they were then qualified to be given a processing machine and a cold room. In August 2017, the processing industry revamped its production from 800 L to 2200 L per day, increasing to 4000 L to date.

The continuous capacity building, in terms of facilitation of the industry staff in the knowledge of marketing, processing technology, exhibitions, and learning visits to other industries both within Tanzania and in the East Africa region, proved to be effective to achieve success.

9.2.3 COMPETITIVE ADVANTAGE

Some of the Grande Demam industry's advantages are the combined expertise of its management team, modernization of production processes, and focus on sales and marketing. The project also took advantage of innovations in technology to expedite its work. It is the opinion of management that the industry's lead time in use of technology will promote project productivity, market penetration, and development of new markets.

In addition, the industry strives to improve quality and packaging of its products using modern technology. Its closeness to market in northern Tanzania also gives it a competitive edge over its competitors, such as Tanga Fresh from Tanga region (about 438 km distant) and ASAS from Iringa in the Southern Highlands zone (701 km distant). Besides, it aims at developing management database systems, which allows the industry to effectively interface data on products and market information with existing customers and supply management personnel. This will assist them to reach areas of problem identification, product features refinement, and real-time solutions to challenges, especially in marketing.

However, the major competitive strength of Grande Demam over others, besides closeness to market, is the quality of products, especially yogurt, a variety of cheeses, and butter, as well as competitive prices for its range of products.

Processing technology used for making Grande Demam products adheres to global formulas for the products to meet international standards.

Impacts of the intervention by MIVARF include:

- Improved economic status of about 1,980 contact farmers
- Improved community health with access to processed milk
- Purchase of refrigerated trucks, improving product sales and marketing
- Mobilization and training of dairy farmers on environmental conservation through zero grazing annually
- Increased gender-sensitive employment, especially for the women and youth in the project area

The outcome results of the MIVARF project intervention are:

- An estimated 2,500 farmers have been contracted to supply milk to the collection stations. It is assumed that with good pricing and availability of cooler tanks, farmers would supply an average 30,000 L of milk twice a week. This goal will be realized in less than six months.
- More importantly, milk losses have been reduced from 80% to nearly 5%.

- Farmers will be assured of access to modern dairy extension services.
- All the food supplements and veterinary medicines are made available at reasonable prices at the industry's milk collection stations on credit terms, and all the costs are deducted from milk sales.
- The milk market will be sustainable, assuring a continuous source of income to the farmers between 95,000 Tanzanian shillings per month from a single animal to 450,000 Tanzanian shillings per month from more animals.

In terms of revenue generation, the project will translate into significant revenue turnover with the lowest small-scale farmer in rural areas earning more than the current Tanzanian minimal wage per month. On the high revenue scale, the farmers will earn an equivalent salary similar to that of a middle-class employee.

In line with the current government vision of industrial development, the MIVARF support has contributed to significant impact(s). As milk flows out of the communities' services for investment, and dairy production flows into the communities, a sustainable cash flow necessary for poverty reduction, rural development, and a better life for the rural community is ensured.

9.3   CONCLUSION

MIVARF trainings in postharvest technologies and provision of matching grants have enhanced produce value addition, reduced crop losses, and improved overall produce prices through quality, safety, and marketing improvement. Value addition also opened accessibility to regional and out-of-region markets, thus making producer groups fetch higher produce prices, improve food security for their farm families, and eventual loss reduction.

REFERENCES

Abass, A. B., Ndunguru, G., Mamiro, P., Alenkhe, B., Mlingi, N., and Bekunda, M. (2014). Postharvest food losses in a maize-based farming system of semi-arid savannah area of Tanzania. *Journal of Stored Products Research*, 57, 49–57.

Abdoulaye, T., Ainembabazi, J. H., Alexander, C., Baributsa, D., Kadjo, D., Moussa, B., Omotilewa O., Ricker-Gilbert, J., and Shiferaw, F. (2016). Postharvest loss of maize and grain legumes in Sub-Saharan Africa: Insights from household survey data in seven countries. PURDUE Extension/Agricultural Economics. Expert Reviewed. EC-807-W. cited on 16/8/2016

Barreiro-Hurle, J. (2012). Analysis of incentives and disincentives for maize in the United Republic of Tanzania. Technical notes series, MAFAP, FAO, Rome, Italy.

Diei-Ouadi, Y. and Mgawe, Y. (2011). Postharvest fish loss assessment in small scale fisheries. A guide for extension officer. FAO Fisheries and Aquaculture Technical paper 559. Rome, Italy.

MIVARF. (2017). Outcome study report.

Southern Agricultural Growth Corridor of Tanzania (SAGCOT). (2010). Appendix IV: value chain and market analysis. Southern Agricultural Growth Corridor of Tanzania. SAGCOT uploads.

URT. (2016a). Agricultural sector development programme I (ASDP I). Support through basket fund. The United Republic of Tanzania. Government Programme Document.

URT. (2016b). Agricultural sector development programme II (ASDP II). Support through basket fund. The United Republic of Tanzania. Government Programme Document.

Section III

Trainings

10 Global Postharvest E-Learning Programs via the Postharvest Education Foundation

Devon Zagory and Deirdre Holcroft

CONTENTS

10.1 INTRODUCTION

By 2040, the world's population is projected to reach nine billion. Feeding this population will be a daunting task. Considerable effort is focused on improving crop yields, breeding more efficient varieties, and better use of resources. But much of the food that is currently produced in the developing world is lost through postharvest deterioration and insect infestations, while a substantial amount goes to waste in the developed world by discarding edible food. Reducing postharvest loss (PHL) and waste is the most resource efficient way to increase the food supply now, and in the future. Many agencies, large and small, public and private, governmental and non-governmental organizations (NGOs), are working to address the problem.

The Postharvest Education Foundation (PEF) was founded in 2011 by Dr. Lisa Kitinoja, and established with the assistance of a small group of like-minded colleagues, with an aim to provide the motivation, training, and mentoring for postharvest professionals around the world. PEF works with a wide clientele, including trainers in NGOs, horticulture companies, extension workers, research scientists, postharvest professionals, and graduate students in Africa, Latin America, South Asia, Southeast Asia, and the Middle East. PEF's mission is to provide innovative programs that motivate and empower people to reduce food losses and waste. At the heart of the PEF strategy is a structured e-learning program for professionals in the developing world interested in learning hands-on practices that directly address PHL. E-learning is an effective way to reach practicing scientists and extension agents in distant places that may otherwise be difficult to reach. It is a cost-effective way to disseminate information and experience among people of differing levels of expertise. Keeping costs low enables PEF to train and mentor a large number of candidates, and those who complete the program can then go on to extend that training to their own regions. This chain reaction has been effective in delivering maximum impact with minimum inputs.

10.2   FOOD LOSSES AND WASTE

Recently the issues of global food losses and food waste have been in the news, with estimates that 30%–40% of food produced goes to waste before it can be consumed. Global food losses and waste (sometimes referred to as FLW) vary widely depending upon the type of food, and occur during production, postharvest handling, food processing, storage, distribution, retail, and consumption (Gustavsson et al. 2011). PHL, which occur earlier in the handling chain, are more severe in developing countries. The losses may be due to inadequate infrastructure, poor temperature management, low levels of technology, and/or low investment in the food production systems, especially the cold chain.

Key factors affecting food losses include:

- Lack of understanding of harvest indices of plant foods and how maturity is related to quality and shelf life
- Poor sorting and grading practices during preparation for market, allowing damaged and decaying foods to enter the supply chain and spread decay
- Inadequate temperature management and lack of control of relative humidity, leading to shriveling, wilting, and deterioration of perishable foods
- Packaging that provides poor protection during handling, transport, and storage
- Delays in transport to market without proper storage (cool storage for perishables, drying of staple grains/beans/legumes before storage)
- Lack of appropriate postharvest handling practices and technologies, leading to rough handling, mechanical damage, improperly handled mixed loads, and contamination concerns
- Lack of the utilization of sustainable, cost-effective postharvest practices, leading to high levels of food loss on the farm, and in wholesale and retail markets
- Lack of storage, food preservation, and marketing options
- Inadequate education and training.

Food waste is more of a problem in industrialized countries, due in part to retailers and consumers throwing away perfectly edible foodstuffs. Per capita waste by consumers in Europe and North America is 95 kg–115 kg/year, while in sub-Saharan Africa and South and Southeast Asia only 6 kg–11 kg/year is discarded by consumers (Gustavsson et al. 2011).

Key factors affecting food waste include:

- Sorting and grading standards on the farm and in the packinghouse, which have more to do with appearance (color, size, shape) than nutritional value or eating quality, leading to higher discards of edible foods
- Packages and packaging materials based on cosmetic features rather than on strength, cleanliness, ventilation, moisture control, or decay prevention
- Overreliance on long-term cold storage, refrigeration, and freezing or improper storage, resulting in off flavors, chilling injury and freezer burn, and increasing waste along the supply chain
- Confusing or unnecessary "sell-by" or "use-by" dates, based upon cosmetic changes or inventory management schemes, rather than on food safety or quality concerns, leading to waste of edible foods at the retail level
- Oversized portions served in restaurants, leading to uneaten food being discarded
- Lack of education regarding proper packaging, cooling/cold storage, storage of cooked foods, and reusing leftover foods, leading to increased discards of foods in the home.

Education and training can help people reduce food loss and food waste. But training in postharvest handling is critically lacking in developing countries. Several reputable short courses and workshop programs are offered globally including Postharvest Technology of Horticultural Crops at the University of California, Davis (United States), and Postharvest Technology at Wageningen University (The Netherlands). More recently a year-long International Online Course on Postharvest and Fresh-Cut Technologies was launched by Universidad Politécnica de Cartagena (Spain) which provides a more accessible training program, although focused more on developed countries. There are, however, many people who do not have the financial resources required to travel internationally to these workshops or even to participate in online courses.

In order to reach a wider audience, particularly those who have been unable to participate in these traditional postharvest educational programs, PEF has developed a series of assignments that can provide training in situ. Feedback on the assignments is provided via email, web-based chat, or, more recently, via social media.

10.3 THE POSTHARVEST EDUCATION FOUNDATION

PEF (http://postharvest.org/), is a non-profit organization based in Oregon, United States, whose mission is to provide innovative programs that motivate and empower people to reduce FLW. PEF provides information, advice, training, and mentoring of young professionals who are involved in the fields of agriculture, horticulture, home economics, and food processing to reduce food losses on farms, in markets, and in villages throughout the developing world.

The goals of PEF are to:

- Conduct a variety of postharvest e-learning programs for young professionals who work with small-scale farmers in developing countries
- Provide free access to postharvest training materials for those who are involved in extension work and training of farmers, produce handlers, small-scale food processors, and marketers
- Supply postharvest tools and basic equipment for use in applied research and for improving practical field operations
- Organize postharvest workshops for e-learners who successfully complete their online programs
- Offer long-term mentoring for participants in e-learning programs via social networking websites
- Conduct short courses, study tours, and workshops
- Provide advice and guidance for establishing local postharvest training and services centers

10.4 PEF'S GLOBAL POSTHARVEST E-LEARNING PROGRAM

10.4.1 HISTORY

The PEF Global Postharvest E-learning Program was developed based on instructional materials designed by Dr. Lisa Kitinoja in the 1990s for extension agents in Egypt, and modified for users in Jordan, Lebanon, Morocco, India, Indonesia, and many African countries between 2001 and 2010. The training program was originally developed for young horticultural professionals, thus many of the readings and examples focus on the improved postharvest handling of fruit or vegetable crops. Over past 7 years, the e-learning program has been expanded by PEF to include researchers, extension workers, NGO staff, and graduate students who are dealing with other crops such as grains, legumes, pulses, root and tuber crops, coffee, tree nuts, herbs, and spices.

From 2011 to 2016 the Global Postharvest E-learning Program was offered as a mentor-guided learning program. Participants enrolled in January and worked through ten assignments at their own pace. This allowed those working full-time or undertaking graduate studies to participate and complete the program by the end of December. Since 2011, the program has been managed by a voluntary Board of Directors at PEF, who provide feedback and mentoring. The program was funded by participant registration fees (US$300) and private donations to PEF.

As of November 2016, the PEF Postharvest E-learning Manual was posted online so individuals or groups can participate at no cost, and on their own schedules (Kitinoja 2016). The agenda provided in the manual is similar to that of the 2016 PEF Global Postharvest E-Learning Program, but is provided without a fixed calendar. This course includes self-exams and access to a free online open forum for individuals to receive feedback and encouragement as they move through the assignments at their own pace. Postharvest toolkits and closing workshops are optional components, and require a budget to cover the costs (US$400 for a basic tool kit; approximately US$2,000 per person for attending a closing workshop).

10.4.2 COURSE CONTENT

The first assignment in the course is a comprehensive survey of the background, skill set, and postharvest experience of the student to provide a baseline measure of their knowledge, skills, and experience. The final assignment repeats this survey to determine progress. The survey asks participants to rate their knowledge and experience in:

- General horticultural production information
- General understanding of postharvest technology
- Needs assessment and program development
- Teamwork
- Training and teaching experience
- Specific horticultural technology topic areas

Training materials and supporting documents for each assignment are loaded into a Google drive shared folder (Kitinoja 2016). Participants are provided with a comprehensive list of postharvest resources in an early assignment and are encouraged to join the social media sites associated with PEF, including discussion forums on LinkedIn.com, Twitter, and Facebook (see the PEF website for details). Conversations are conducted via Skype or Facebook Messenger. A list of useful videos available on YouTube and other sites is provided. These can be used to clarify topics or used as training aids in outreach by the trainees.

The heart of the e-learning program is a Commodity Systems Assessment Methodology (CSAM) that leads students through a methodology to identify causes and sources of PHL and quality problems. Each participant completes this assessment on the crop(s) and postharvest handling system(s) of interest to the participant. This assignment guides the students through the 26 components of the assessment using a manual that has been developed and refined over many years (Figure 10.1) (LaGra 1990; LaGra et al. 2016). The end products of CSAM encompass both traditional loss assessment and cost-benefit analysis, and lead to productive extension program and project development by helping identify the causes and sources of PH, and prioritize research, extension, and advocacy needs for the crop (Tables 10.1 and 10.2).

The CSAM manual includes sample data collection methods and detailed explanations of each of the components. Ideally, teams of people work together while investigating a commodity system. For example, a horticultural production researcher might be teamed with a marketing specialist and an extension agent. CSAM can help build links between agencies and individuals, close information gaps, and help people to solve problems while focusing on usable postharvest technology.

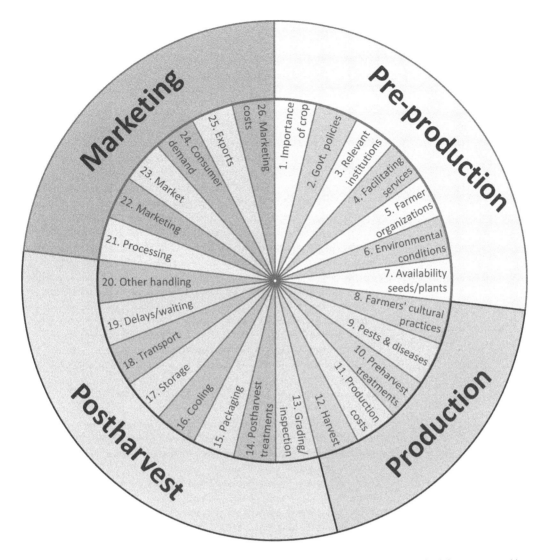

FIGURE 10.1 The 26 principal components of the commodity systems assessment methodology presented in a figure. (Redrawn from LaGra, J. et al., *Commodity Systems Assessment Methodology for Value Chain Problem and Project Identification: A First Step in Food Loss Reduction*, IICA, San Jose, Costa Rica, pp. 246, 2016.)

The manual is a detailed document, complete with sample questionnaires for each component. It includes examples for many different crops and lists of questions and ideas for creating locally specific assessments. It also provides many examples of how to organize and present data. The suitability of "best postharvest practices" and appropriate technologies for the community and clientele are determined. Trainees then conduct a cost/benefit analysis using one or more improved practices and technologies (Tables 10.2 and 10.3).

A key goal of the PEF program is to train the trainers, and consequently the assignments included will help trainers in becoming familiar with small-scale postharvest handling practices and appropriate technologies; designing postharvest demonstrations for local farmers, traders, processors, and marketers; setting measurable goals and objectives for a postharvest training program; and using postharvest extension methods, simple postharvest tools, and basic equipment for quality assessment and as training aids. The assignments

TABLE 10.1

A Commodity Systems Assessment Report on Cassava (*Manihot esculenta*) Grown in Ghana. Report Submitted as Part of the E-Learning Program

Component	Assessment of Cassava
A. Pre-processing	
1 Importance of cassava	Cassava is the most important tuber crop in Ghana, accounting for 30% of daily calorie intake and a per capita consumption of 152.9 kg/year. About 70% of farmers grow cassava on 886,000 ha and produce 15 million metric tonnes.
2 Governmental policies	Some support from the government in developing improved varieties, however many farmers find it difficult to access these improved planting materials. No policy support from the government in terms of subsidies or marketing.
3 Relevant institutions	Council for Scientific and Industrial Research (CSIR) Food Research Institute, CSIR Crop Research Institute, CSIR Soil Research Institute, CSIR Savanna Agriculture Research Institute, Ministry of Food and Agriculture, Bill and Melinda Gates Foundation, University of Ghana–Legon, Kwame Nkrumah University of Science and Technology–Kumasi, Agricultural Cooperative Development International (ACDI)/ Volunteers in Overseas Cooperative Assistance (VOCA), The Export Trade, Agricultural and Industrial Development Fund (EDAIF). Food and Agriculture Organization, United States Agency for International Development (USAID), Deutsche Gesellschaft für Internationale Zusammenarbeit (GIZ), and Agriculture Development Bank.
4 Facilitating services	Access to agrochemicals, extension services, and a road network is available to the majority of farmers. Tractor powered ploughing services are limited, and only about 10% of cassava farmers plough their lands. Credit is limited. The Agricultural Development Bank is one of the few banks providing limited credit to farmers under rigorous conditions. Money lenders and middlemen charge exorbitant interest rates. There are no direct subsidies from the government to support cassava farmers.
5 Producer/shipper organizations	There are no nationwide cassava producer and shipper organizations, but local producers are sometimes organized on a small scale.
6 Environmental conditions	Cassava grows in most soils with a slightly acidic (pH 4.5–6.5), loamy soil and good drainage. Average temperatures of 25°C–29°C and 1,000–1,500 mm per annum of evenly distributed precipitation are optimal.
7 Availability of planting material	Government and international organizations have spent large sums of money in the development and release of high-yielding cassava varieties. These varieties should be readily available from all district offices of the Ministry of Food and Agriculture in cassava growing areas. However, many farmers are not able to get sufficient stock of these improved varieties.
B. Production	
8 Cultural practices	Cassava yields best when the soil is ploughed, harrowed, and ridged, but many farmers plant directly after clearing the land without ploughing or ridging. Cassava should be planted in rows at a spacing of 1 × 1 m. Many farmers fail to optimize plant density and, consequently, yield. Weed control is manual and should be done at least three times during production. Disease control is done by applying chemicals, or by destroying virus-infected plants.
9 Pests and diseases	Diseases attacking cassava plants include cassava mosaic virus (leaf mottling); cassava bacterial blight (angular leaf spots); brown leaf spot (round, brown spots with definite border on the upper surface of mostly older leaves); cassava white leaf spot (white spots with a red border); cassava fungal blight (large spreading brown leaf spots mainly on leaf tips and margins); cassava root rot (soft and decaying tubers with offensive odor). Major pests of cassava are rats, cane rats (grasscutters), grasshoppers, and mice.

(*Continued*)

TABLE 10.1 (*Continued*)
A Commodity Systems Assessment Report on Cassava (*Manihot esculenta*) Grown in Ghana. Report Submitted as Part of the E-Learning Program

	Component	Assessment of Cassava
10	Preharvest treatments	Weeding before harvesting is highly recommended since it facilitates harvesting, especially for mechanical harvest.
		Irrigating the soil prior to harvest is recommended, especially in the dry season when the ground is hard.
11	Production costs	Production costs are provided in Table 10.2.

C. Postharvest

	Component	Assessment of Cassava
12	Harvest	Harvesting is usually done manually, preferably when the soil is moist.
		Cassava harvesting equipment is available.
		Depending on the variety, cassava is harvested 10–18 months after planting.
		Cassava harvested after 18–24 months have a lower starch content and a higher incidence of decay.
		Cassava is transported in baskets, headpans, or sacks.
13	Grading, sorting, and inspection	No grading is done, although decayed tubers are removed.
		Inspection is not mandatory and is not usually performed.
14	Postharvest treatments	Curing, which would be beneficial in reducing losses after harvest, is not practiced.
		Cassava tubers are highly perishable and can last for about three days, after which the starchy flesh begins to brown and weight loss occurs.
15	Packaging	Tubers are either packed in sacks and loaded onto trucks or loaded directly onto trucks. This packaging is not suitable because it results in bruising and rapid spoilage.
		Shelf life can be extended by packing tubers free of wounds and bruises in small boxes (<10 kg).
16	Cooling	None.
17	Storage	Cassava is highly perishable and can be stored for about 3 days without any intervention.
		Postharvest losses of fresh cassava roots could be as much as 100% and result from poor harvesting, handling, and transportation.
		Cassava tubers are often stored by heaping them on the floor and covering them with wet jute sacks or by placing them in containers containing water. Sometimes, the tubers are placed in either moist trenches in the ground or placed in moist saw dust.
		Usually harvested tubers are held at temperatures of 26°C–35°C and 50%–60% relative humidity (RH). If tubers are stored at 2°C–5°C and 90%–95% RH, they can last for more than 2 months.
18	Transport	Cassava tubers are transported in baskets, headpans, or sacks from farms to the local processing or marketing centers.
		Small vehicles and tricycles are also used to transport the tubers from farms to market or to local processing centers.
		Trucks are used to transport the tubers from farm gates or village market centers to distant cities.
		Tubers are loaded and unloaded from the trucks haphazardly without due diligence. This damages the tubers.
19	Delay/Waiting	Delays are not typical.
20	Other operations	There is sufficient labor, but workers are untrained, and they do not harvest and handle the product carefully.

(*Continued*)

TABLE 10.1 (*Continued*)

A Commodity Systems Assessment Report on Cassava (*Manihot esculenta*) Grown in Ghana. Report Submitted as Part of the E-Learning Program

Component	Assessment of Cassava
D. Processing	
21 Agroprocessing	Cassava can be processed into various products including gari (roasted cassava granules or grits), agbelima (cassava dough/wet mash), fufu (pounded boiled cassava), high quality cassava flour (HQCF), industrial grade cassava flour (IGCF), high quality cassava chips (HQCC), starch, ethanol, glucose syrup, and tapioca.
	The equipment for processing into starch and HQCF is not readily available in Ghana, which limits this as an option for farmers. The support of the CAVA II project is increasing the demand for starch, HQCF, and IGCF cassava products.
	Producing gari or agbelima is more common since these products do not require much capital for a startup.
22 Marketing intermediaries	Cassava tubers can be sold directly to consumers or processors, but more typically they go through various intermediaries before reaching the wholesaler, retailer, exporter, or processor.
	Intermediaries finance the farmers and control production and price.
	Large-scale processors contract growers to provide cassava at a competitive price.
	Losses within the chain are borne by whoever is in charge of the tubers at the time of losses.
23 Market information	Mobile phones have made marketing information readily available, and there are several mobile marketing service providers including Esoko, Farmerline, and Farm Radio International. These mobile marketing service providers provide timely, accurate, and reliable information on prices of farm produce at various locations.
	However, there are many farmers and marketers who are unaware of these services, do not have mobile phones, or cannot understand the language used by the service providers. Ghana has many languages and only a few are used by these services.
24 Consumer demand	Local consumers expect fresh, unwaxed product, but do not have a size preference.
	Exporters prefer medium-sized, straight tubers with a surface factor of about 1.2.
	Processors, especially starch and HQCF processors, prefer fresh tubers because any delay in processing reduces the starch content.
	During the rainy season (June to October) there is usually an oversupply of tubers, while supply can be limited during the dry season (December to May).
25 Exports	Export demand is low.
26 Marketing costs	Cassava is either sold at the farm gate or transported to local markets.
	Cost of transportation is usually the responsibility of the wholesaler or the aggregator.

Source: Dramani, Y., Commodity systems assessment on cassava (*Manihot esculenta*), Unpublished report, 2016.

include designing postharvest training and extension programs for various audiences, as well as evaluating the effectiveness of these programs.

An optional assignment included in the training is that of designing a Postharvest Training and Services Center (PTSC) for the trainee's community based on successfully completed projects (Kitinoja and Barrett 2015). This assignment includes budgets and is intended to serve as a proposal to be submitted to governments or funding agencies.

TABLE 10.2

Production Costs of Two Different Methods of Cassava Production in Ghana

Input or Activity	Quantity	Unit Cost (GH¢)	Total Cost (GH¢)	US$ ($1= GH¢3.75)
Method 1: Total Yield = 16 t/ha				
Land Rent	1 ha	250	250	66.67
Land preparation: clearing		15	225	60.00
Planting: planting material	100 bundles	free		
Planting: labor	10 work days	15	150	40.00
Weeding: manual	30 work days	15	450	120.00
Harvesting	5 work days	15	75	20.00
Total cost of production			**1,150.00**	**306.70**
Method 2: Total Yield = 45 t/ha				
Land Rent	1 ha	250	250	66.67
Land preparation: ploughing	2 times	250	500	133.33
Land preparation: herbicide cost	5 L	13	65	17.33
Land preparation: herbicide application	3 work days	15	45	12.00
Planting: improved planting material	100 bundles	2	200	53.33
Planting: labor	10 work days	15	150	40.00
Weeding: manual	30 work days	15	450	120.00
Weeding: herbicide cost	3 L	13	39	10.40
Weeding: herbicide application	3 work days	15	45	12.00
Harvesting	10 work days	15	150	40.00
Total cost of production			**1,894.00**	**505.07**

Comparisons of the two different production methods and their relative benefit:cost ratios:

Comparing Production Methods

Method	Production Cost (GH¢/ha)	Yield (t/ha)	Price (GH¢/t)	Total Income (GH¢)	Net Income (GH¢)	Benefit: Cost Ratio
1	1,150	16.9	200	3,380	2,230	1.9
2	1,894	45.0	200	9,000	7,106	3.8

Source: Dramani, Y., Commodity systems assessment on cassava (*Manihot esculenta*), Unpublished report, 2016.

10.4.3 COMPLETION AND CLOSING WORKSHOP

Those trainees who successfully complete all the e-learning assignments and submit four written reports receive a Postharvest Train-the-Trainer Certificate of Completion, and can purchase a Postharvest Tool Kit valued at over US$400 at a subsidized price of US$300. The toolkit contains a refractometer, digital scale, digital temperature probe, measuring tools, color charts, quality rating scales, quality assessment tools, and an assortment of postharvest training aids and supplies.

Successful trainees are invited to travel at their own expense to work with our PEF team for a closing workshop on implementing training programs for small-scale horticultural farmers at one of many new PTSCs. The top three participants (as evaluated and selected by the training team and PEF board of directors) are awarded a PEF travel grant valued at up to US$1,500 to cover travel expenses to attend the workshop.

TABLE 10.3

Postharvest Cost/Benefit Example Worksheet. In This Example, the Harvest is 1000 kg of Sweet Peppers and the Current Packaging (Sacks) Is Compared to Reusable Plastic Crates (New Practice). All Costs Are in US$ (Kitinoja 2016)

Item	Current Practice	New Practice
Describe:	Sacks for sweet peppers 25 kg	Reusable plastic crates plus fiberboard liners 12.5 kg
Costs:		
40 sacks @ $0.50	$20	
80 crates @ $6.00		$480
Crates liners @ $0.10		$8
Relative cost	$20	$488
Recurring costs	$20	$8
Expected benefits:		
% losses	30%	5%
Amount for sale	700 kg	950 kg
Value/kg	$1.00/kg	$1.25/kg
Total market value	$700.00	$1,187.50
(Market value) − (Costs) =	$680.00	$699.50
Relativeprofit first load		$19.50
Relative profit second and subsequent loads		$499.50
Return on investment		**1 load of 1000 kg**

Source: Kitinoja, L., PEF postharvest e-learning manual training of postharvest trainers and extension specialists: Small-scale postharvest handling practices and improved technologies for reducing food losses, 2016.

Notes: In this example, even though the cost for plastic crates is very high, the new practice is immediately profitable with the first load. Once the plastic crates have been paid for, the relative profit will rise to +$499.50 per load of 1000 kg. Plastic crates can be reused more than 100 times.

The closing workshop is usually organized in conjunction with a local university or another conference. The program includes presentations from PEF board members, local professors, and PEF e-learners as well as site visits and demonstrations.

In addition, each year the PEF Board confers the Kader Award to the PEF e-learning program graduate who has had the greatest impact in providing education to local farmers, traders, processors, and/or marketers, and helping them to reduce food losses in their village, country, or region after the successful completion of the course. The award honors the late Professor Adel A. Kader, who we believe contributed more to the knowledge and understanding of postharvest technology of fruits and vegetables than any other individual in the field. Dr. Kader's knowledge, experience, and generosity of spirit continue to inspire us today through PEF and the Kader Awards. The award includes a certificate in honor of Dr. Kader, a trip to a PEF-sponsored event, and a US$500 cash prize.

10.4.4 TIMEFRAME

The PEF Global E-Learning Program is typically offered over 8 to 14 months. Programs taking longer or shorter periods of time have been implemented over the years, depending upon the type of project and number of e-learners in the group. One year has been found to be sufficient time for participants to absorb the enormous amount of information on postharvest technology and extension education methods provided during the program, to interact with one another across many time zones via online forums, and to develop confidence in their new knowledge and skills.

10.4.5 Participants and Impact

Since 2011, 149 individuals from 31 countries, predominantly in South Asia and Africa, have graduated from the Global E-Learning Program (Table 10.4), and 35 individuals registered for 2017. These graduates provide an annual report on their progress and, over the last 4 years, their training events (including training of trainers) and outreach efforts have reached nearly 120,000 small farmers, handlers of horticultural crops, and extension agents.

TABLE 10.4
Geographical Spread and Number of Graduates of the PEF Global E-Learning Program as of December 2016

Country	Graduates
Bangladesh	3
Benin	2
Bhutan	7
Botswana	1
Cambodia	4
Cameroon	3
Chile	1
Egypt	1
Ethiopia	15
Germany	1
Ghana	12
India	8
Indonesia	1
Iran	1
Kenya	13
Lebanon	1
Malawi	1
Malta	1
Namibia	1
Nepal	2
Nigeria	5
Pakistan	5
Peru	1
South Africa	1
Rwanda	9
Sri Lanka	1
Tanzania	25
Togo	2
Uganda	7
United States	12
Zambia	2
Total	**149**

10.5   CONCLUSIONS

The PEF Global Postharvest E-Learning Programs have been operating for over 7 years. The graduates regularly communicate with the PEF and have had success in training and outreach. Their expertise in identifying the causes of food losses, their knowledge and use of postharvest technologies such as solar drying, zero energy cool chambers, processing, packaging, storage, and marketing, as well as postharvest training skills have furthered their impact and their careers. Several work as private consultants, while others have jobs as postharvest researchers, extension officers, trainers, and educators. All of these graduates have played an important role in reducing food losses in their communities.

REFERENCES

Dramani, Y. (2016). Commodity systems assessment on cassava (*Manihot esculenta*). Unpublished report.

Gustavsson, J., Cederberg, C., Sonesson, U., van Otterdijk, R., and Meybeck, A. (2011). *Global Food Losses and Food Waste - Extent, Causes and Prevention*. FAO, Rome, Italy. Retrieved from http://www.fao.org/docrep/014/mb060e/mb060e00.pdf.

Kitinoja, L. (2016). PEF postharvest e-learning manual training of postharvest trainers and extension specialists: Small-scale postharvest handling practices and improved technologies for reducing food losses. Retrieved from http://www.postharvest.org/PEF%20Training%20of%20Postharvest%20Trainers%20Manual%202016%20FINAL.pdf

Kitinoja, L. and Barrett, D.M. (2015). Extension of small-scale postharvest horticulture technologies–A model training and services center. *Agriculture*, 5, 441–455.

LaGra, J. (1990). A commodity systems assessment methodology for problem and project identification. Moscow, ID, Postharvest Institute for Perishables, University of Idaho, IICA, AFHB, pp. 102.

LaGra, J., Kitinoja L. and Alpizar, K. (2016). *Commodity Systems Assessment Methodology for Value Chain Problem and Project Identification: A First Step in Food Loss Reduction*. San Jose, Costa Rica: IICA. 246 pp. Retrieved from http://repiica.iica.int/docs/B4232i/B4232i.pdf (English) and http://repiica.iica.int/docs/B4231e/B4231e.pdf (Spanish).

11 Creation of Postharvest and Food Processing Videos for Extension of Postharvest Technologies

Diane M. Barrett

CONTENTS

11.1 INTRODUCTION

11.1.1 ADVANTAGES OF VIDEOS OVER OTHER EDUCATIONAL FORMATS

Videos are an extremely fast and effective means of conveying a new idea. Regardless of the language spoken or the country a student lives in, a picture or video is "worth a thousand words." The availability of mobile phones and advances in technology, even in relatively poor countries, have allowed short videos to be viewed easily.

For many years, educators in the United States and around the world have utilized short courses of 1–5 days to educate professionals working full-time in the fruit and vegetable industry. As a Cooperative Extension Specialist at the University of California, Davis, it was obvious in recent years that attendance was declining due to travel costs and time restrictions. The industry is requesting more online education that does not require its employees to leave their jobs. The development of short, targeted videos to supplement distance programs is highly desirable.

11.1.2   The Effectiveness of Videos in the Conduct of Postharvest Education

Videos have been used since the early 2000s to record classroom instruction, visits from guest lecturers, explanations of difficult problems, and other supplemental information to students (Kay 2012). In most instances, viewing is done in a passive manner; for example, the viewer watches the video in a remote location from the speaker. In a review of over 53 videos created between 2003 and 2011, Kay (2012) found numerous positive benefits when students viewed videos. These included more control over learning, improved student habits, and increased learning performance.

These videos do not require a subject matter specialist to feature them to the audience. Hence, relatively more farmers with different educational levels can benefit from the videos shown by organizations, with little previous agricultural experience (Bentley et al. 2014).

11.1.3   Current Use of Videos in Postharvest Technology Education

While educational videos are being incorporated into numerous educational platforms, relatively few of them focus on the best methods for postharvest handling and preservation of fruits and vegetables. For the postharvest audience, videos with instructions about quantification of losses in both yield (weight) and quality, as well as how to process fruits and vegetables to extend shelf life, would be very useful. The Postharvest Technology Center (PTC) at the University of California (UC), Davis, has an excellent video library at this link: http://postharvest.ucdavis.edu/libraries/video/. Most of the videos listed are also posted on the following YouTube channel: http://www.youtube.com/user/ucdpostharvest.

These videos are organized under seven general topic areas:

* Harvest and postharvest handling systems
* Produce food safety
* Handling and nutrition
* Sensory Evaluation
* Small-scale postharvest handling
* Transportation
* Flowers and ornamentals

If they are short enough, videos may be used during the PTC's short courses, or they may be viewed independently. This video library page also lists the year when the video was created. In the past, videos were often created in a format of a longer classroom lecture, with slides and narration; however, more recent videos may be shorter in length (4 minutes–20 minutes) and feature a headshot of the speaker and video.

The UC Davis International Programs Office, which exists under the College of Agriculture, has created many short YouTube videos for extension teaching activities as part of their e-Afghan program, available on their YouTube channel. These videos are less than six minutes long, are intended for remote audiences, and are formatted with simple text overlaid on the video, and music in the background. The use of text rather than narration allows for easy translation into other languages, and the video itself is understandable in all languages.

The Postharvest Education Foundation (PEF) also maintains a video library on their YouTube channel at this link: https://www.youtube.com/channel/UCgPycz8ZVEwj4vWgaZyIgig. These videos include slide presentations from postharvest short courses and relevant videos from organizations such as Food Tank, the World Vegetable Center, Archer Daniels Midland (ADM), and others.

Another academic group called Scientific Animations Without Borders (SAWBO), based at the University of Illinois, uses animation and narration for their very effective videos, which can be viewed at http://sawbo-illinois.org/main.htm. Some of the SAWBO videos do address postharvest issues, such as postharvest grain storage, hand washing, and disease prevention. Their detailed activities are discussed in another chapter of this book.

11.2 CREATION OF A DATABASE OF EXISTING VIDEOS

A video search was conducted prior to starting to create new videos, in order to determine what was already available, and in what format. In the fall of 2013, over 60 videos were viewed, including those from the UC Davis PTC and other resources mentioned above. The topics could be grouped under the following three headings:

1. Fruit and vegetable postharvest
2. Fruit and vegetable quality evaluation
3. Fruit and vegetable preservation

A Microsoft Excel database of the existing videos was created, which included columns for the following terms: subject matter, creator, website address, short description, scale and audience, video type (video, narration, text, and music), length, and year produced. A large number of the videos reviewed had either narration only, or perhaps narration overlaid on the video. Creators of these videos kept them to 10 minutes in length for the most part, but some were more lecture style at 30–60 minutes. Not all the videos were included, only those that were the most informative and not overly commercial. This Microsoft Excel database is available at the link: http://www.fruitandvegetable.ucdavis.edu/Fruit_-_Vegetable_Videos/.

11.3 EXAMPLES OF EXISTING VIDEOS

Links to the over 60 videos in the database of fruit and vegetable videos are available and are organized into the three topic areas mentioned above. An overview of each topic, as well as examples of videos in each of these areas, are mentioned in the following section.

11.3.1 Fruit and Vegetable Postharvest Handling

There are 33 videos related to the postharvest handling of fruits and vegetables located at the link above. Table 11.1 illustrates an example of how the videos appear in the Microsoft Excel database.

TABLE 11.1

Examples of Videos on the Topic of Postharvest Handling of Fruits and Vegetables in the Database

Subject Matter	Creator	Website Address	Target Audience	Video Type	Length	Year
Small-Scale Postharvest Practices	UC Davis, PTC	http://www.youtube.com/ watch?v=i2GQJx99yU4	Small- or medium-scale	Slides and narration	9:28	2008
Postharvest management of fruits & vegetables	Indira Gandhi National Open University, India	http://www.youtube.com/ watch?v=95RRdoySdjA	Small- or medium-scale	Video, music, narration	21:45	2009
Postharvest Loss Prevention: Dry, Inspect, Clean and Examine (D.I.C.E.)	Scientific Animations Without Borders, University of Illinois	https://sawbo-animations. org/video.php?video=// www.youtube.com/embed/ f27uHCnQoS0	Small- or medium-scale; farmer, trader	Animation	5:02	2018

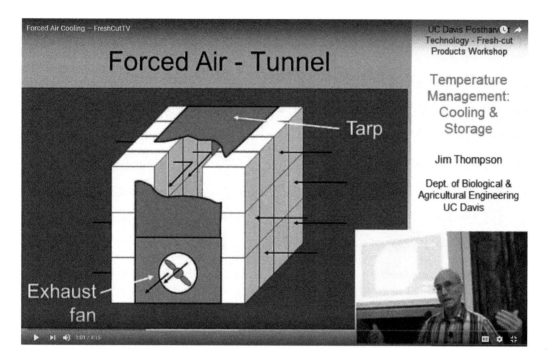

FIGURE 11.1 Screenshot of forced air cooling video.

Lectures from a number of short courses put on by the PTC at UC Davis have been videotaped and posted online. One example is a video on forced air cooling by Jim Thompson, an emeritus Agricultural Engineer: https://www.youtube.com/watch?v=0SfoIbclGUA&feature=youtu.be. This video includes a headshot of the speaker, as well as the slides presented, and was made with the software Camtasia. Methods of temperature management are described, and instructions on how to assemble a forced air cooler with a fan to evacuate air in the middle of stacked crates of produce are reviewed (Figure 11.1).

11.3.2 Fruit and Vegetable Quality Evaluation

At the time that the database was created, there were only seven videos on fruit and vegetable quality evaluation that were deemed of appropriate content and length. This points out the need for more videos on this topic. A short list from this part of the database appears in Table 11.2.

At UC Davis, the International Programs Office has created excellent videos with music and text overlaid, which describe methods of agricultural production, as well as a few methods for evaluation of fruit and vegetable quality. "How to Use a Refractometer" is one example that may be viewed at the following link: https://afghanag.ucdavis.edu/video/refractometer. Step-by-step illustrations of how to use a refractometer are given. Fruits or vegetables are squeezed through cheesecloth to extract the juice, and the percentage of Brix (soluble solids) is read off a grid, then the refractometer is cleaned and dried until the next use (Figure 11.2).

11.3.3 Fruit and Vegetable Preservation

Under the topic of fruit and vegetable preservation, there were almost 40 videos in the Microsoft Excel database. Some examples from various organizations appear in Table 11.3. While many of these videos were created by universities or educational organizations, there are also many created by commercial businesses who sell equipment or ingredients used for preservation.

TABLE 11.2

Examples of Videos Related to Fruit and Vegetable Quality Evaluation in the Database

Subject Matter	Creator	Website Address	Target Audience	Video Type	Length	Year
Use of a Refractometer	UC Davis International Programs Office	https://afghanag.ucdavis.edu/video/refractometer	Medium- or large-scale	Video & text	1:25	2013
Using a Psychrometer - Measuring Relative Humidity	UC Davis International Programs Office	https://afghanag.ucdavis.edu/video/psychrometer -measuring-relative-humidity-video	Medium- or large-scale	Video & text	2:08	2013
How to use a fruit hardness tester (Penemetro para fruta)	InfoAgro (Infoagro.com)	http://www.dailymotion.com/video/xb0s09_how-to-use -a-fruit-hardness-tester_tech	Medium- or large-scale	Video & text	0:58	2009

FIGURE 11.2 Refractometer video with an illustration of the application of juice to the prism.

One tremendous resource is the United States National Center for Home Food Preservation (NCHFP), which has many videos on home canning, freezing, and dehydrating on its website: http://nchfp.uga.edu/multimedia_videos.html#video. All of these excellent videos are less than three minutes and describe safe methods of food preservation. "Hot pack for vegetables" for example, provides a step-by-step description of the process used for canning vegetables with a pH > 4.6.

TABLE 11.3

Examples of Videos Available to Highlight Methods of Fruit and Vegetable Preservation

Subject Matter	Creator	Website Address	Scale	Video Type	Length	Year
Canning at different altitudes	National Center for Home Food Preservation	http://nchfp.uga.edu/multimedia_videos.html#video	Home/small	Illustrations and text	0:28	1990
How to Remove the Poison from Cassava Flour	Scientific Animations Without Borders, University of Illinois	https://www.youtube.com/watch?v=p_JqS9kwueI	Home/small	Animation	2:48	2014
Juice - Safer Processing	California Department of Public Health, United States Food and Drug Administration, Centers for Disease Control, UC Davis	http://ucfoodsafety.ucdavis.edu/Food_Processing/Fruit_and_Vegetable_Juices/	Medium/large	Interviews	2:36	2010

11.4   PRIORITIZED NEEDS FOR NEW VIDEOS IN THE POSTHARVEST AREA

Videos with a short, simple format using overlaid text were felt to be the easiest to translate and most likely to be understood by the largest variety of audiences. Short videos of less than six minutes appeared to be the most popular, therefore this length was chosen. In addition, a decision was also made to prioritize "how-to" videos, which described how to use the various tools that a postharvest scientist would need to conduct evaluations. Of the three topic areas reviewed (e.g., fruit/vegetable postharvest handling, quality evaluation, and preservation), the fewest videos available were on the topic of quality evaluation.

Therefore, "How to Use a Color Chart to Increase Market Value" was the first video created in 2013, and it is posted at the following YouTube channel: http://www.youtube.com/channel/UC4Z7Gf-Tf8UsUkxyu54PeKw. A "Technical Note" that accompanies the video includes the objectives, key concepts, materials required, background material, how learning will be reinforced/evaluated, references, and discussion questions. This note is available here: http://www.fruitandvegetable.ucdavis.edu/Fruit_-_Vegetable_Videos/. Since it was posted in 2014, the color video has been viewed over 500 times, indicating that it does meet an industry need.

PEF carries out an e-learning program (http://www.postharvest.org/postharvest_elearning_program1.aspx), after which many of the graduates receive a "postharvest toolkit" with tools that can be used to measure postharvest losses in quantity and quality. Creation of a "How to Use…" video for each tool would allow postharvest scientists to easily understand its use (Barrett 2014).

In 2016, two additional videos were created, one on "How to use a refractometer" (viewed over 6,300 times) and another on "How to measure temperature and relative humidity" (viewed over 250 times to date).

11.5   THE SEQUENCE OF STEPS TO BE FOLLOWED IN MAKING TECHNICAL NOTES, NARRATIVE SCRIPTS, AND VIDEOS

1. Selection of audience. Is your audience small-, medium- or large-scale farmers? This will greatly affect the content and format of the video.
2. Identify the key concept. This should be focused, concise, and clear for your particular target group. Both benefits and risks should be identified. The UC Davis International Agriculture

Extension Academy has developed a series of fact sheets to aid in the identification of key concepts. These can be found at https://www.agextonline.com/message-form-and-delivery.html and https://www.agextonline.com/ (U.C. Davis International 2007).

3. Language. Once the target audience has been identified, prepare the materials in a language that is common to most. If the format of the video is simple, with only written words, it should be easy to translate the video to other languages.

4. Technical note. It is usually easier to prepare a short, written document, one or two pages, describing the information you wish to convey prior to starting to create the video itself. Some guidelines for fact sheets have been developed by UC Davis http://www.agextonline.com/uploads/3/2/4/3/3243215/fs_ext_making_fact__sheets.pdf (U.C. Davis International 2013). This technical note might include the following topics:

    a. Objectives
    b. Key concepts
    c. Background information
    d. Materials required
    e. How learning will be reinforced/evaluated
    f. Discussion questions
    g. References

5. Video. The video itself may be created in iMovie or a similar computer program. One may take videos of fruits and vegetables, handling practices, market scenes, etc., to help illustrate the key concepts and other topics covered in the technical note. Tips on making effective videos have been described at the following link: http://www.agextonline.com/uploads/3/2/4/3/3243215/fs_ext_video_shooting_tips.pdf (U.C. Davis International 2012).

    a. Each video clip should be sufficiently long (1–3 minutes) so that the "best" sections can be used for the actual video. Often clips are too short, or images are not clear, therefore it is better to err on the side of too much material.
    b. When demonstrating something in the video, it is usually best to focus on hands if possible, and not include the entire body of the demonstrator, which can be distracting.
    c. Once the video clips are selected, they can be edited and positioned in such a way that they tell a clear story.

6. Narrative script. The script is easiest to add after the video clips have been selected. Focus on using the fewest, simplest words possible, in a font large enough to see clearly.

11.6 TIPS FOR PREPARING EFFECTIVE VIDEOS

1. Importance of Time Management (Length of Video). Most effective videos are less than 6–8 minutes in length. Rather than making a longer video, aim to create a number of short, more focused videos conveying a specific topic.

2. Materials and Resources to be Collected Before Starting the Video. As mentioned above, iMovie is one software program for creating videos, but there are others. It is important to determine the amount of bandwidth available to your target audience so that you do not create a resource that cannot be viewed. Other materials required for postharvest videos include fruits and vegetables, postharvest technologies such as picking bags, returnable plastic crates, or whatever the focus of the video is on.

3. Budget Management. The primary capital costs of creating videos will be the camera, computer, and software program used. In addition, the labor required is a significant part of the budget.

4. Precautions. It is best to start with a very simple, focused topic and create a "trial video" so that one gets accustomed to using the resources and materials involved. Keep the focus very specific and the rewards will be more attainable.

REFERENCES

Barrett, D.M. (2014). Creating fruit and vegetable postharvest technology videos. PEF White Paper No. 14-01. Retrieved from http://postharvest.org/PEF_WhitePaper_Fruit_Vegetable_Videos_DMB_2014%20FINAL.pdf.

Bentley, J., Van Mele, P., Okry, F., and Zossou, E. (2014). Videos that speak for themselves: When non-extensionists show agricultural videos to large audiences. *Development in Practice*, 24(7), 921–929.

Kay, R.H. (2012). Exploring the use of video podcasts in education: A comprehensive review of the literature. *Computers in Human Behavior* 28, 820–831.

U.C. Davis International Ag Extension Academy. (2007). Design and plan your message delivery options. Retrieved from http://www.agextonline.com/message-form-and-delivery.html.

U.C. Davis International Ag Extension Academy. (2013). Make a good fact sheet. Retrieved from http://www.agextonline.com/uploads/3/2/4/3/3243215/fs_ext_making_fact__sheets.pdf.

U.C. Davis International Ag Extension Academy. (2012). Video shooting tips. Retrieved from http://www.agextonline.com/uploads/3/2/4/3/3243215/fs_ext_video_shooting_tips.pdf.

12 Scientific Animation Without Borders (SAWBO)

Addressing Educational Gaps around Value Chains Pertaining to Postharvest Losses

Julia Bello-Bravo, Anne Namatsi Lutomia, and Barry R. Pittendrigh

CONTENTS

12.1 INTRODUCTION

The Food and Agriculture Organization (FAO) estimated that as much of one-third of food produced is lost before people consume it. While food security concerns have inspired a worldwide response (FAO 2011), they are deepened by the fact that the world will have to feed an increasingly large population in the coming decades (Godfray et al. 2010). One major area for improving food security is by reducing postharvest loss (PHL) (Hodges et al. 2011).

PHL typically differs dramatically in developing, as opposed to developed, countries. That is, for developing nations, more potential for reducing PHL occurs towards the beginning of a value chain, amongst the producers and other actors who live, work, and gain their livelihood at the farm or farm gate level. To reach these stakeholders and provide sustainable solutions for food security gaps (United Nations 2016) means leveraging—rather than seeing as a liability—regions often limited in local infrastructure and resources, including low or absence of literacy in national languages.

This requires not only craft and creativity, but recognizing how constraint drives, rather than inhibits, creativity. The constraint of limited or no local print literacy, for instance, suggests film, video, and/or animation as useable media instead. Similarly, the constraint of limited or no local infrastructure for educational content delivery suggests video-enabled cell phones to deliver and redistribute such content to local users (Bello-Bravo et al. 2017a). By taking such constraints as productive (for offering solutions now), rather than as problems (to be circumvented or solved first), then animated, culturally accessible, educational videos supported by an ever-expanding digital superstructure (Aker 2010) represent an economically, socially, and environmentally sustainable format for effectively transferring scientifically grounded knowledge to nearly any population affected by issues of public concern, including PHL.

12.1.1 Cell Phones and Education for Postharvest Loss Prevention

Around the world, an expanding mobile industry has facilitated access to, and the use of, mobile phones, even in rural and remote areas, as a way to enable the flow of information and access to technology services for farmers in those areas (Feder and Savastano 2017; Omotayo 2005; Walter et al. 2017). As discussed in Hudson et al. (2017), African countries continue to lead the world in the growth of the mobile markets, reaching 420 million subscribers by the end of 2016.

As in developed countries, cell phones have changed how people in developing countries live (Aker 2010). For farmers and other production side stakeholders, cell phones afford faster and cheaper communications (Donner 2007), greater access to markets (Bello-Bravo et al. 2017b), access to previously cost prohibitive or geographically distant agricultural extension services (Asongu and Biekpe 2017; Odiaka 2015), and other experiences beyond simple economic livelihood (Ganesh 2010). With these gains, however, also come new challenges for farmers: e.g., gender differences in rural mobile phone access, as well as use and recurrent social inequalities due to the rural/urban divide (Chen and Wellman 2005; Gillwald et al. 2010; Hafkin 2000). Taking these new constraints as productive for PHL solutions, video-enabled cell phones afford a comparatively less expensive option, and generally provide higher access to digital infrastructures that make them well-suited for delivering all types of educational material (whether text, graphics, visual/animated, or audio).

12.1.2 Animated Videos and Education for Postharvest Loss Prevention

Significant numbers of development projects use live action or animated video, or both, as a component of training programs and curricular support (Bello-Bravo and Baoua 2012; Bello-Bravo et al. 2013a; Bello-Bravo and Pittendrigh 2012; Gandhi et al. 2007; Ladeira and Cutrell 2010; Medhi et al. 2012; Munyua 2000). Van Mele (2011) found that approximately 78% of development organizations' studies used videos in their farmer-focused training programs. Bello-Bravo et al. (2017b) and Maredia et al. (2017) as well found that video-based training was at least as, and in some cases more, effective than traditional extension approaches regarding knowledge transfer.

Video-based educational content (VBEC) affords multiple technological advantages, especially for agriculture training in developing countries (Van Mele 2011). With proper audio and visual equipment and adequate space, VBEC can be projected practically anywhere (Coldevin 2003). Cell phones—or other small or handheld ICT devices—can even more affordably provide this needed equipment and adequate space. For example, market women in Ghana watched animated videos on loaned cell phones while conducting business at their market stalls. Cell phones are already more common than the relatively specialized equipment needed for video projection, but cultural mechanisms also enable the sharing of cell phones amongst people without them (James and Versteeg 2007; Wesolowski et al. 2012). Cloud-based access and wireless sharing also makes storage and accessibility of VBEC available virtually everywhere, while the small size of animated video files makes their distribution and redistribution by producers and consumers easier as well.

In terms of technological literacy, already more widespread familiarity with cell phones as a digital technology makes it less frequently necessary, and generally takes less time if necessary, to train potential learners on how to use an app to access VBEC on cell phones (Bello-Bravo et al. 2017b). When such training is necessary for learners completely unfamiliar with electronic technologies, it is generally easier and quicker than training them to the computer literacy needed to access and utilize computer-based software learning (Tata and McNamarra 2018). Moreover, while female extension agents being trained on a farm productivity software package preferred face-to-face interactions over computer-based ones (Tata and McNamarra 2018), women have consistently described the appeal of learning via cell phones (Bello-Bravo and Baoua 2012; Bello-Bravo et al. 2013a, 2017a, 2017c, in press). VBEC also does not require print literacy—a major advantage in rural areas where literacy education is typically underfunded, unequally accessible, especially cross gender lines, or simply not available (Kiramba 2017).

Localizing the verbal component of VBEC in dialectically accurate speech further enhances these advantages (Bello-Bravo et al. 2017b). While connecting VBEC to the actually-lived experiences and interests of learners is essential (Knowles 1984; Vygotsky 2012), to produce animated videos in locally accurate dialects is not just communicative pragmatism (Nonaka and Takeuchi 1997); addressing people in their own language also fosters trust in solutions communicated (Levin and Cross 2004; Szulanski et al. 2004), making them more authoritative and potentially liable to uptake (Bello-Bravo et al. 2017a). Accuracy of dialect, however, is essential. While people from diverse cultural groups are generally open to learning from the same or similar sets of culturally nonspecific animated images, dialectical inaccuracy of voiceovers (the recorded spoken audio associated with the given animation) is reported as off-putting and diminishes the effectiveness of knowledge transfer (Bello-Bravo and Baoua 2012; Bello-Bravo et al. 2013a, 2013b, 2013c, 2017c). Moreover, not using local dialects in communications inhibits participation in even locally fluent people (Kiramba 2017), while an exclusive use of national or ex-colonial languages generally accessible only to elites effectively enables their participation only (Bunyi 2008). Such local dialect use also counters any prevailing rural/urban power dynamics and represents a democratizing step that invites participation by otherwise marginalized community members in a problem that affects the community as a whole. Importantly, this "offset" does not exclude national or ex-colonial language elites from participation, since most are fluent (if not native) in local dialects (Bunyi 2008).

As a medium, VBEC on cell phones generally affords a more economical and culturally attuned vehicle with wider reach than other alternatives, although animated content is simpler and less economically resource intensive to produce than live action content, and more easily adapts linguistically to local dialects. Audiovisual educational content itself, however, whether live action or animated, is also already more effective for knowledge transfer compared to text-only content (Ladeira and Cutrell 2010; Medhi et al. 2012; Moreno and Mayer 2002). Animated videos on cell phones, then, may help to foster effective knowledge transfer (1) by placing the learner in control of the experience (Bello-Bravo et al. In press), (2) by avoiding problems around authority and authoritative speaking on educational topics that sometimes arise between teachers and students (Bello-Bravo et al. 2017b; Jansen 2001), and/or (3) by defusing the sheer power effects of ex-colonial or class-biased languages when used for education (Bourdieu and Passeron 1990; Martin-Jones and Heller 1996).

Despite the capacity of such audiovisual material for explaining complicated concepts, raising awareness, and influencing people's decisions (Lie and Mandler 2009), their real potential exists as an integration into localized educational strategies. The "holy grail" of necessary and sufficient conditions—the "mix"—of educational content and technical/pedagogical delivery (Marton 2018), of learner motivation and state of mind (Knowles 1984), of educational environment and its connection to the learner's world outside that environment (Vygotsky 2012), and how these components interact remains imperfectly determined. Nonetheless, these culturally appealing, digitally accessible, and freely (re)distributable cell phone accessible animated educational videos offer an effective and economically, socially, and environmentally sustainable format that meets Agenda 2030's call for knowledge transfer solutions around food security and PHL (United Nations 2016). As a format, it affords scalable educational and technological interventions at any link along the food production value chain.

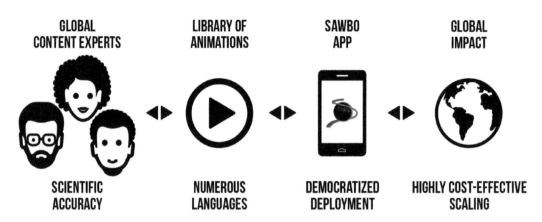

FIGURE 12.1 SAWBO systems approach schematic.

Any PHL value chain solution is itself already the outcome of a production value chain as well. As a primarily intellectual and creative process, these "knowledge chains" are as integrally a component of any effective and sustainable educational solution to PHL as the work product itself. Figure 12.1 depicts the four basic links of a Scientific Animation Without Borders (SAWBO) knowledge chain: (1) identifying and scientifically framing a problem, so that solutions to that framing of the problem are implied, (2) materializing that scientifically abstract solution in a concrete, overdubbed as needed animated video as a means for effecting knowledge transfer about the problem's importance, context, and need to act on its solution, (3) maintaining a video distribution infrastructure that affords educators, learners, and researchers free and easy access to SAWBO videos, and provides SAWBO a means for collecting and incorporating feedback on any video or part of its knowledge chains itself, and (4) watching for ways to more widely scale up the reach of any video and the processes that developed it.

Along these knowledge chains, SAWBO works hyper-collaboratively as described by Kessler 2013. Concretely, this means heuristically leveraging the wide reach and speed of current digital infrastructures not only to network with otherwise inaccessible global, local, and indigenous knowledge holders, but also to very rapidly and iteratively update any developing knowledge/data pools around a given knowledge link in the chain. More abstractly, this means maintaining a type of openness and responsiveness to feedback that contrasts both with traditional command strategies and more recent open innovation network and partner ecosystem strategies (Radjou and Prabhu 2015). If projects, institutions, and so forth collaboratively "seek external know-how, they are disposed toward knowledge that reinforces rather than challenges their worldview" (Radjou and Prabhu 2015), then hyper-collaboration, by contrast, welcomes, even invites, challenges. This can serve to better establish "matters of fact" (Cook 2007) about a problem or the processes at work along a knowledge chain. Analogous to the scientific method, it rests on a validated empiricism (Charmaz 2014; Cook 2007; Glaser 2002) that grounds learning and can enable not just shifts of behavior but shifts of paradigm as well (Wals 2015). The flexibility, adaptiveness, and resilience (Folke et al. 2002; Gunderson 1999) that this fosters manifests in the range of creative means and methods used by SAWBO, not only at the level of creating the final animated work product but also at the level of choosing approaches for accomplishing that goal in the first place. Such social, learning-oriented capacity building is what Wals (2015) emphasizes as indispensable for generating sustainable solutions and strategies for problems at the project, institutional, community, national, or global levels.

Like PHL in developing countries, where interventions at the top of the value chain afford the greatest rewards but also the greatest risks, the beginning of a SAWBO knowledge chain similarly affords the greatest rewards and risks. This, not only because it is vitally necessary to identify the problem in such a way that all of the downstream work done on solutions actually address how the problem is lived by people in the world, but also because of the need to identify and assemble a requisite variety of knowledge-holders (local, global, and indigenous) to identify the problem in the first place.

Once tackled, feedback between the developed/developing knowledge pool and the video construction team centers on three key issues: (1) what empirically valid approach has a broad community consensus for framing and solving the problem (i.e., an idea already beyond the conceptual only or test phase stage), (2) how can this be effectively conveyed in a brief series of understandable steps in a video, and (3) how will affected stakeholders gain easy and affordable access to information or technology required by the solution in the video?

At the third link, hyper-collaboration through digital infrastructures especially affords access to local, dialectically fluent language-holders even in very remote places, not only to transmit back any voiceovers recorded on cell phones, but also to recommend script changes, especially when terminology in the video requires better adaptation to local vocabulary and examples. Similarly, however, once an animation has been released, it represents "cat out of the bag" information that cannot be taken back or easily "untaught" in target communities. Although any created educational content will ideally represent the community knowledge-holders' agreed upon best practices approach as a solution, the flexibility, adaptiveness, and resilience of the video creation knowledge chain also allows for updates in light of feedback, even long after the project is officially completed.

As with PHL in developing countries, where interventions at the end of the chain afford the least reward but also the least risk, the end of a SAWBO knowledge chain similarly marks the "easiest" step. To date, 12 SAWBO animations have specifically addressed various PHL prevention applications. These are freely available through digital download (http://sawbo-animations.org), on portable USB libraries, via the SAWBO Deployer App for cell phones (https://goo.gl/Hs2EVZ), and in higher resolution files for mass media broadcasting. While considerable research has addressed PHL due to insect or disease degradation (Pimentel and Raman 2003; Sosa et al. 2016), losses during transport (Mohammed et al. 2016), and other value chain problems (Affognon et al. 2015), less has been done to date on PHL solutions related to markets (FAO 2011; Kaminski and Christiaensen 2014). While the SAWBO animations focused on this part of the value chain typically emphasize less-empowered entrepreneurs in regions—including women, rural producers, and those with little to no formal education—each embodies a scalable, flexible, adaptive, and resilient solution to an identified value chain gap. In the following, then, we describe three cases of applied knowledge chain solutions for preventing PHL.

12.2 THREE CASE STUDIES AS APPLICATIONS OF SAWBO KNOWLEDGE CHAINS

12.2.1 Case Study #1: Framing the Problem: Processing Shea Nuts for Shea Butter in Nigeria

Shea butter is a vegetable oil produced from the African shea tree (*Villetaria paradoxa*) in at least 18 African countries, including Nigeria (Maranz and Wiesman 2003). Having a wide variety of indigenous uses traditionally, over the past century shea butter has grown into a multibillion U.S. dollar per year world commodity. Also, traditionally, its production has largely been the work of, and a major source of income for, women (Elias and Carney 2005). Despite this potential boon, however, numerous value chain gaps affect PHL for shea butter production—most visibly in the inability of traditional production methods to keep pace with world demand. While classic developmentalism might advance modernization and production optimization to address these gaps, not only is this economically unfeasible for women, who have fewer financial resources than men (Elias 2003), it also would not align with Agenda 2030's call for more economically, socially, and environmentally sustainable food security solutions (United Nations 2016).

Modernized farming methods would not only more severely impact local soil quality and the environment compared to traditional methods of shea tree cultivation (Bello-Bravo et al. 2015), it would affect the social landscape as well. Increasing the gross production of shea trees in Nigeria would not guarantee increased raw materials to women for the production of shea butter. Although women traditionally have had access to, and the use of, shea fruit (Elias 2003), the ability to practice this right remains subject to approval by local elites and/or male landowners. If the local owners of trees decide to sell their crop, rather than recognize women's right to its use, women have little social capital to resist this decision. Similarly, developmental productivity recommendations that result in less access to labor for women would not only risk increased poverty and a reproduction of existing social inequalities (Bourdieu 1993), it would even more disparately affects women's social structures in general. Women traditionally work not only individually and cooperatively to produce shea butter, but intergenerationally as well, transmitting information and techniques from older to younger producers apprentice-wise (Bello-Bravo et al. 2015). That local elites and males also would likely resist any perceived challenge or diminishment of their power from women's empowerment also operates as a constraint to any proposed solution here. Many of these constraint details were not apparent when SAWBO began working with others to frame solutions around the problem of PHL for Nigerian shea butter production, but only emerged later hyper-collaboratively.

Having learned that the lipolytic activity of seed lipases and microorganisms is halted by heating and can be desiccated to the needed <8% moisture content (Louppe 1994) for shea butter production, this suggested a possible improvement that would increase yields and reduce early value chain PHL of raw materials without negatively impacting the social world of women's work. This "matter of fact" (Cook 2007) also represented a scientifically-based, teachable content translatable into an animated video, subsequently liable to sharing and redistribution through the area's already available digital infrastructure. As a productivity increase that did not negatively affect women's access to labor or their wider social world, its potential increase of social capital also might enable resistance to prevailing gender inequalities. As a solution, it echoes the finding by Shackleton et al. (2011) that adding value at the process level to environmental products with external markets is key for helping poor families to gain economic and social mobility. That it leveraged, rather than eroded, women's cooperative social structures links it also to their current use of unions as a way to access markets offering higher value, fair trade shea butter. Lastly, as a change to the production chain requiring less firewood and water, its smaller carbon footprint and lower contribution to global climate change also makes it more environmentally sustainable.

In response to this well-defined educational step that has the potential to solve a problem early in the value chain of shea production, SAWBO worked with Dr. Peter Lovett, a global expert on shea, to produce the 2 minute and 20 second animation entitled "Improved Method of Shea Butter Processing" (SAWBO 2016; http://goo.gl/zSp7XA). Describing the steps for improved yields, lighter workloads, and social learning via communities of practice (Wenger 1998) also informed field deployment of the video. Communities of practice reflect ways of doing things, have views, values, power relations, and ways of talking that both organize and make group learning possible. Both informally and formally, this includes people who share a concern or passion about a topic or set of problems, and seek to deepen their understanding of it by interacting with others on an ongoing basis (Wenger 1998). The cooperative and intergenerational social structures of women in this case reflect a community of practice that supports a wider distribution of the social learning afforded by sharing videos on improved shea butter production (Lutomia and Bello-Bravo 2017).

12.2.2 Case Study #2: Democratizing Access: Hermetic Sealing of Market Legumes Using a "Jerrican" Technique in Mozambique

Approximately 73.4%, or 18.4 million people, in Mozambique have mobile subscriptions (CIA 2017). And while agriculture composes the mainstay of Mozambique's economy—with more than one-fourth of the nation's Gross Domestic Product (GDP) and employing 80% of the labor force—the overwhelming majority of these subsistence farmers (99.65%) are smallholders farming an average 1.2 ha of land; another 0.33% are medium-holders with 6.6 ha, and 0.07% as large-holders with an average 282 ha (Mole 2006). Though they grow food for their own consumption, especially grains and beans and/or peanuts, productivity levels for smallholders remain low due to little modernization; after 30 years of independence, most peasant farms have not changed (Mole 2006). Food security is tenuous.

For this case—a study conducted in Zambézia Province, Mozambique by a team from two United States-based universities (Mocumbe et al. 2016) as part of a multidisciplinary research project funded by the United States Agency for International Development (USAID) and its Feed the Future initiative (Mazur et al. 2013)—a total of 314 male and female bean-growing smallholders from two different administrative posts in rural Zambézia province were randomly assigned to one of three experimental treatments: (1) learning only via traditional extension teaching; (2) learning only via animated video on a smartphone; or (3) learning by a combination of these. The curriculum presented a sealed (jerrican) container method for more securely storing dry beans postharvest. Prior interviews had shown that farmers suffered serious bean losses due to bruchid infestation postharvest and that few knew of, or how, to store beans in metal or heavyweight plastic airtight containers to reduce or eliminate PHL. Importantly, rural Mozambican farmers had been involved as knowledge-holders for pretesting the animated video storyboards to be used and provided feedback during the video creation process that changed the final work product.

Quantitatively comparing pre/posttreatment knowledge scores, none of the three groups differed significantly in knowledge about jerrican storage methods prior to the treatments. All three groups (98%) also said they would use this storage method in the future (Mocumbe 2016). While all three groups showed significantly increased posttest knowledge scores, the animation only and animation plus extension treatments showed significantly higher learning compared to extension only (Mocumbe 2016). Separated by sex, however, women's scores significantly improved in the animation only treatment, while men's did not. Moreover, while women pretested significantly lower than men prior to treatments, their posttest knowledge scores were no longer significantly lower than men's (Mocumbe 2016). While animation only and animations plus extension teaching enabled women to "catch up" with men despite starting significantly behind (Mocumbe 2016), it remains unclear what mechanism helped to close this gendered knowledge gap.

Qualitatively, participants engaged more with the format than the content on a first viewing; they giggled at the animation and were impressed to hear their local language. Watching the video a second time, they were more engaged in the content. While controlling the video's pace and viewing enhances knowledge transfer, it also facilitates individualized and meaningful participation by learners. This affords both the time and the space to reflect on and connect any informational content delivered to participants' lived experiences (Vygotsky 2012), and thus helps to democratize access to education. More dramatically, one ad hoc group of individuals who watched the video three times volunteered to demonstrate its content. A male and female participant went to the front of the crowd with beans and a jerrican and began performing what they had learned, interacting with the crowd playfully throughout. Participants in the crowd would offer suggestions from the video as others checked to confirm that the instructions had been carried out properly. During this process, women were not afraid to speak up and contributed, in contrast to at the beginning of the program, when they did not say much.

This collective, conversational, bidirectional "classroom" differed in virtually every way from the traditional extension teaching, with its authoritative, active teacher and a quiet, passive group of students. Instead of vertical classroom dynamics (Kiramba 2017) presented in an elite language that carries heavy markers of real and symbolic power (Bourdieu 1991; Bourdieu and Passerson 1990), participants meaningfully participated in the content of the demonstration in their own language, demanded accountability and accuracy from those offering the demonstration, were recognized, and were comfortable to transgress what initially seemed gendered boundaries. While the procedural legitimacy (Hemmati 2002; Parkinson 2006) of this interaction was wholly unanticipated by the researchers, that such substantive participation denotes a key component of democracy (Manin et al. 1999) suggests that the "classroom" observed in this case was strikingly democratized.

12.2.3   CASE STUDY #3: PIVOTING TOWARDS CHALLENGE: HERMETIC STORAGE
TO PREVENT INVENTORY PHL FOR MARKET-WOMEN IN GHANA

This research—focused on eliminating inventory PHL for legume market women in Ghana (Bello-Bravo et al. 2019)—generally reprised the approach and findings of the previous case. In market settings with unhygienic floors, especially when it rains, that attract pests and rodents, and have few to no raised storage pallets, the traditional use of non-airtight jute bags (or sometimes just open containers) readily leads to PHL from rot, fungus growth, and pest predation but could be remediated by an improved airtight (jerrican) storage method. Offering this genuine alternative, the main purpose of this study, however, was to assess the pre/posttest learning efficacy and appeal of smartphone-delivered animated video content as a cost-effective, accessible information delivery protocol. Its design neither called for nor used a control to test the efficacy of the video itself, partly because previous research had done so (Mocumbe et al. 2016).

As previously, a majority of participants understood the training content after viewing the video, could follow the technique described, liked that the message was in their local language, and expressed interest in trying the jerrican technique (Bello-Bravo et al. 2019). Importantly, participants connected how the prevailing forms of storage influenced inventory losses and recognized that training via smartphone-delivered, animated videos could be relevant to them. Most had limited to no prior experience with video-enabled smartphones, and what few cell phones were owned by participants were viewed as strictly for communication, not for education.

Participants in this study—all of them either market women or farmers who also marketed legumes—were identified by a local agent and SAWBO colleague. Such a locally respected individual is often a literal necessity when working in Africa, to gain access to participants. Despite this purposeful selection (Ritchie et al. 2013), the experimental procedure itself, including interviews by the researchers, was performed while the market women conducted business at their stalls. Loaned a smartphone and briefly instructed on how to play the animations, the market women were free to watch the video at their convenience and as many times as they wished. Several paused mid-video to attend to customers, stopped to chat with friends or take care of their children, who were in the shop with them, or simply set it aside for a time. Notably, smartphones proved pedagogically "small" enough to fit into the small shops of a Ghanaian market woman's busy workday.

While findings for this case generally confirmed its research questions—that participants showed increased learning on the video topic presented and found cell phone learning appealing—they also challenged the offered solution, citing cost (or simply how to acquire jerricans), problems with fitting the non-collapsible jerricans into available market spaces, no need for alternative storage, and, in one case, doubt that jerricans would work. A solution-based implementation too confident in its own knowledge might take these as technical details only or a failure on the part of some participants to "get it." Hyper-collaboratively, these objections challenge the value chain PHL solution and highlight spots where still more exactly framing the problem might have pre-solved these issues: e.g., a locally available, variably sized, airtight and watertight storage container, for instance, would have exactly addressed the environmental liabilities that current jute bags experience without raising concerns about storage space, even if affordability remained an issue. Nonetheless, just as market women recognized new affordances in smartphones that were not previously relevant to them, the possibility of an alternative to traditional storage methods using locally available containers was observed. Consequently, the SAWBO animated video was adjusted to include a diverse range of locally available and affordable container technologies that could be used for hermetic sealing of grains (http://goo.gl/XxkRvP).

12.3   CONCLUSION

Moving forward, it will be critical for the PHL community and SAWBO to continue to determine a common canon of PHL solutions where teachable steps are the critical gap for reducing PHL, identify steps in value chains where these PHL educational solutions can be delivered, determine the needs for localization (both in terms of languages and in some cases the alteration of visuals), and determine the pathways by which these solutions can be delivered to the end users. SAWBO continues to explore approaches for educators to easily obtain and deliver SAWBO animations in the field.

In their "small" way, animated video on cell phones can fit into tightly constrained economic, social, and environmental spaces and times threatened by food insecurity and PHL, whether these are precarious markets, remote backwaters, gender roles, busy workdays, or resource-straitened circumstances. As a vehicle, this format readily delivers the kind of flexible, adaptive, and resiliently sustainable solutions (Wals 2015) to PHL problems that Agenda 2030 (United Nations 2016) both calls for and requires.

REFERENCES

Affognon, H., Mutungi, C., Sanginga, P., and Borgemeister, C. (2015). Unpacking postharvest losses in Sub-Saharan Africa: A meta-analysis. *World Development*, 66, 49–68.

Aker, J. C. (2010). Dial "A" for agriculture: Using information and communication technologies for agricultural extension in developing countries (Working Paper 269). Washington, DC: CGD.

Asongu, S. A., and Biekpe, N. (2017). *Government Quality Determinants of ICT Adoption in Sub-Saharan Africa*. Yaoundé, Cameroon: African Governance and Development Institute.

Bello-Bravo, J., and Baoua, I. (2012). Animated videos as a learning tool in developing nations: A pilot study of three animations in Maradi and surrounding areas in Niger. *The Electronic Journal of Information Systems in Developing Countries*, 55(6), 1–12.

Bello-Bravo, J., Dannon, E., Agunbiade, T., Tamo, M., and Pittendrigh, B. R. (2013a). The prospect of animated videos in agriculture and health: A case study in Benin. *International Journal of Education and Development using Information and Communication Technology*, 9(3), 4–16.

Bello-Bravo, J., Dannon, E. A., Zakari, O. A., Amadou, L., Baoua, I., Tamò, M., and Pittendrigh, B. R. (2017c). An assessment of localized animated educational videos (LAV) versus traditional extension presentations or LAV followed by extension agent discussions among farmers in Benin and Niger [Abstract]. Presented at The Feed the Future Legume Innovation Lab: Grain Legume Research Conference, Ouagadougou, Burkina Faso, 13–18 August 2017.

Bello-Bravo, J., Lovett, P. N., and Pittendrigh, B. R. (2015). The evolution of shea butter's "paradox of paradoxa" and the potential opportunity for information and communication technology (ICT) to improve quality, market access and women's livelihoods across rural Africa. *Sustainability*, 7(5), 5752–5772.

Bello-Bravo, J., Lutomia, A. N., Madela, L. M., and Pittendrigh, B. R. (2017a). Malaria prevention and treatment using educational animations: A case study in Kakamega County, Kenya. *International Journal of Education and Development using Information and Communication Technology*, 13(1), 70–86.

Bello-Bravo, J., Lutomia, A. N., and Pittendrigh, B. R. (2019). Changing formal and informal learning practices using smartphones: The case of market women of Ghana. In D. P. Peltz and A. C. Clemons (Eds.), *Multicultural Andragogy for Transformative Learning* (pp. 171–193). Hershey, PA: IGI Global.

Bello-Bravo, J., Olana, G. W., Enyadne, L. G., and Pittendrigh, B. R. (2013b). Scientific animations without borders and communities of practice: Promotion and pilot deployment of educational materials for low literate learners around adama (Ethiopia) by Adama Science and Technology University. *The Electronic Journal of Information Systems in Developing Countries*, 56.

Bello-Bravo, J., and Pittendrigh, B. R. (2012). Scientific Animations Without Borders: A new approach to capture, preserve and share indigenous knowledge. *The Journal of World Universities Forum*, 5(2), 11–20.

Bello-Bravo, J., Seufferheld, F., Agunbiade, T., Steele, L. D., Guillot, D., Ba, M.,... Pittendrigh, B. R. (2013c). Scientific Animations Without Borders™: Cell-phone videos for cowpea farmers. Presented at the Innovative research along the cowpea value chain (Proceedings of the 5th World Cowpea Research Conference), Saly, Senegal, 26 September–1 October 2010.

Bello-Bravo, J., Tamò, M., Dannon, E. A., and Pittendrigh, B. R. (2017b). An assessment of learning gains from educational animated videos versus traditional extension presentations among farmers in Benin. *Information Technology for Development*, 1–21.

Bourdieu, P. (1991). *Language and Symbolic Power*. Cambridge, MA: Harvard University Press.

Bourdieu, P. (1993). *The Field of Cultural Production: Essays on Art and Literature*. New York: Columbia University Press.

Bourdieu, P., and Passeron, J. C. (1990). *Reproduction in Education, Society and Culture*. Cambridge, MA: Harvard University Press.

Bunyi, G. (2008). *Constructing Elites in Kenya: Implications for Classroom Language Practices in Africa Encyclopedia of Language and Education* (pp. 899–909). New York: Springer.

Charmaz, K. (2014). *Constructing Grounded Theory*. Thousand Oaks, CA: Sage Publications.

Chen, W., and Wellman, B. (2005). *Charting Digital Divides: Comparing Socioeconomic, Gender, Life Stage, and Rural-Urban Internet Access and Use in Five Countries Transforming Enterprise: The Economic and Social Implications of Information Technology*. (pp. 467–497). Cambridge, MA: MIT Press.

CIA. (2017). *World Factbook: Mozambique*. Retrieved from https://www.cia.gov/library/publications/the-world-factbook/geos/mz.html.

Coldevin, G. (2003). *Participatory Communication: A Key to Rural Learning Systems. Communication for Development*. Rome, Italy: FAO. Retrieved from http://www.fao.org/nr/com/gtzworkshop/Participatory%20communication%20a%20key%20to%20rural%20learning%20systems.pdf.

Cook, H. J. (2007). *Matters of Exchange: Commerce, Medicine, and Science in the Dutch Golden Age*. New Haven, CT: Yale University Press.

Donner, J. (2007). The rules of beeping: Exchanging messages via intentional "missed calls" on mobile phones. *Journal of Computer-Mediated Communication*, 13(1), 1–22.

Elias, M. S. (2003). Globalization and female production of African shea butter in Rural Burkina Faso. (MSc), University of California Los Angeles, Los Angeles, CA.

Elias, M. S., and Carney, J. (2005). Shea butter, globalization, and women of Burkina Faso. In J. Seager and L. Nelson (Eds.), *A Companion to Feminist Geography* (pp. 93–108). Oxford, UK: Basil Blackwell.

FAO. (2011). *Global Food Losses and Food Waste-Extent, Causes and Prevention*. Rome, Italy: FAO. Retrieved from http://www.fao.org/fileadmin/user_upload/suistainability/pdf/Global_Food_Losses_and_Food_Waste.pdf

Feder, G., and Savastano, S. (2017). Modern 2 agricultural technology adoption in Sub-Saharan Africa. *Agriculture and Rural Development in a Globalizing World: Challenges and Opportunities*, 11.

Folke, C., Carpenter, S., Elmqvist, T., Gunderson, L., Holling, C. S., and Walker, B. (2002). Resilience and sustainable development: Building adaptive capacity in a world of transformations. *AMBIO: A Journal of the Human Environment*, 31(5), 437–440.

Gandhi, R., Veeraraghavan, R., Toyama, K., and Ramprasad, V. (2007). Digital green: Participatory video for agricultural extension. Paper presented at the Information and Communication Technologies and Development, 2007.

Ganesh, I. M. (2010). 'Mobile love videos make me feel healthy': Rethinking ICTs for development. *IDS Working Papers*, 2010(352), 1–43.

Gillwald, A., Milek, A., and Stork, C. (2010). *Towards Evidence-based ICT Policy and Regulation: Gender Assessment of ICT Access and Usage in Africa*. Cape Town, RSA: Research ICT Africa.

Glaser, B. G. (2002). Constructivist grounded theory? Paper presented at the Forum qualitative sozialforschung/forum: Qualitative social research.

Godfray, H. C., Beddington, J. R., Crute, I. R., Haddad, L., Lawrence, D., Muir, J. F., Pretty, J., Robinson, S., Thomas, S. M., and Toulmin C. (2010). Food security: The challenge of feeding 9 billion people. *Science*, 327(5967), 812–818.

Gunderson, L. (1999). Resilience, flexibility and adaptive management—Antidotes for spurious certitude? *Conservation Ecology*, 3(1). Retrieved from https://www.consecol.org/vol3/iss1/art7/.

Hafkin, N. (2000). Convergence of concepts: Gender and ICTs in Africa. In E. M. Rathgeber and E. O. Adera (Eds.), *Gender and the Information Revolution in Africa* (pp. 1–15). Nairobi, Kenya: International Development Research Centre.

Hemmati, M. (2002). *Multi-Stakeholder Processes for Governance and Sustainability: Beyond Deadlock and Conflict*. Sterling, VA: EarthScan.

Hodges, R. J., Buzby, J. C., and Bennett, B. (2011). Postharvest losses and waste in developed and less developed countries: Opportunities to improve resource use. *The Journal of Agricultural Science*, 149(S1), 37–45.

Hudson, H. E., Leclair, M., Pelletier, B., and Sullivan, B. (2017). Using radio and interactive ICTs to improve food security among smallholder farmers in Sub-Saharan Africa. *Telecommunications Policy*, 41(7/8), 670–684.

James, J., and Versteeg, M. (2007). Mobile phones in Africa: How much do we really know? *Social Indicators Research*, 84(1), 117.

Jansen, J. D. (2001). Image-ining teachers: Policy images and teacher identity in South African classrooms. *South African Journal of Education*, 21(4), 242–246.

Kaminski, J., and Christiaensen, L. (2014). Post-harvest loss in Sub-Saharan Africa—What do farmers say? *Global Food Security*, 3(3–4), 149–158. doi:10.1016/j.gfs.2014.10.002

Kessler, G. (2013). Teaching ESL/EFL in a World of Social Media, Mash-Ups, and Hyper-Collaboration. *TESOL Journal*, 4(4), 615–632.

Kiramba, L. K. (2017). Translanguaging in the writing of emergent multilinguals. *International Multilingual Research Journal*, 11(2), 115–130.

Knowles, M. S. (1984). *Andragogy in Action*. San Francisco, CA: Jossey-Bass.

Ladeira, I., and Cutrell, E. (2010). Teaching with storytelling: An investigation of narrative videos for skills training. *Presented at the Proceedings of the 4th ACM/IEEE International Conference on Information and Communication Technologies and Development*, London, UK, 2010.

Levin, D. Z., and Cross, R. (2004). The strength of weak ties you can trust: The mediating role of trust in effective knowledge transfer. *Management Science*, 50(11), 1477–1490.

Lie, R., and Mandler, A. (2009). *Video in Development: Filming for Rural Change*. Wageningen, the Netherlands: CTA.

Louppe, D. (1994). *Le Karité en Côte d'Ivoire*. Nogent-sur-Marne, France: CIRAD-Forêt.

Lutomia, A. N., and Bello-Bravo, J. (2017). Communities of practice and indigenous knowledge: A case study of empowering women in processing shea butter using scientific animations. In *Handbook of Research on Social, Cultural, and Educational Considerations of Indigenous Knowledge in Developing Countries* (pp. 226–243): IGI Global. Hershey, PA: Information Science Reference.

Manin, B., Przeworski, A., and Stokes, S. (1999). Introduction. In B. Manin, A. Przeworski and S. Stokes (Eds.), *Democracy, Accountability, and Representation* (pp. 1–26). Cambridge, UK: Cambridge University Press.

Maranz, S., and Wiesman, Z. (2003). Evidence for indigenous selection and distribution of the shea tree, Vitellaria paradoxa, and its potential significance to prevailing parkland savanna tree patterns in Sub-Saharan Africa north of the equator. *Journal of Biogeography*, 30(10), 1505–1516.

Maredia, M. K., Reyes, B., Ba, M. N., Dabire, C. L., Pittendrigh, B. R., and Bello-Bravo, J. (2017). Can mobile phone-based animated videos induce learning and technology adoption among low literate farmers? A field experiment in Burkina Faso. *Information Technology for Development*, 1–32. doi:10.1080/02681102.2017.1312245.

Martin-Jones, M., and Heller, M. (1996). Introduction to the special issues on education in multilingual settings: Discourse, identities, and power: Part I: Constructing legitimacy. *Linguistics and Education*, 8(1), 3–16.

Marton, F. (2018). Towards a pedagogical theory of learning. In K. Matushita (Ed.), *Deep Active Learning* (pp. 59–77). Singapore: Springer.

Mazur, R., Abbott, E., Lennssen, A., Luvaga, E., Yost, Y., Bello-Bravo, J.,,... Maria, R. (2013). *Farmer Decision Making Strategies for Improved Soil Fertility Management in Maize-Bean Production Systems*. East Lansing, MI: Feed The Future.

Medhi, I., Menon, R. S., Cutrell, E., and Toyama, K. (2012). Correlation between limited education and transfer of learning. *Information Technologies and International Development*, 8(2), pp. 51–65.

Mocumbe, S. (2016). Use of animated videos through mobile phones to enhance agricultural knowledge among bean farmers in Gurúè District, Mozambique. Master of Arts (MA), Iowa State University, Ames, IA.

Mocumbe, S., Abbott, E., Mazur, R., Bello-Bravo, J., and Pittendrigh, B. R. (2016). Use of animated videos through mobile phones to enhance agricultural knowledge and adoption among bean farmers in Gúruè District, Mozambique. *Presented at the Association for International Agricultural and Extension Education*, Portland, OR, April 4–8, 2016.

Mohammed, M., Wilson, L. A., and Gomes, P. I. (2016). Sodium hypochlorite combined with calcium chloride and modified atmosphere packaging reduce postharvest losses of hot pepper. *International Journal of Research and Scientific Innovation*. 32, pp. 1–9. Retrieved from. http://www.rsisinternational.org/IJRSI/Issue32/01-09.pdf

Mole, P. N. (2006). *Smallholder Agricultural Intensification in Africa* (Mozambique Micro Study Report). Maputo, Mozambique: AFRINT.

Moreno, R., and Mayer, R. E. (2002). Verbal redundancy in multimedia learning: When reading helps listening. *Journal of Education Psychology*, 94(1), 156–163.

Munyua, H. (2000). Information and communication technologies for rural development and food security: Lessons from field experiences in developing countries. Rome, Italy: FAO.

Nonaka, I., and Takeuchi, H. (1997). *The Knowledge-Creating Company*. Oxford, UK: Oxford University Press.

Odiaka, E. C. (2015). Perception of the influence of home videos on youth farmers in makurdi, Nigeria. *Journal of Agricultural And Food Information*, 16(4), 337–346.

Omotayo, O. (2005). ICT and agricultural extension. Emerging issues in transferring agricultural technology in developing countries. Agricultural Extension in Nigeria Ilorin: Agricultural Extension Society of Nigeria.

Parkinson, J. (2006). *Deliberating in the Real World: Problems of Legitimacy in Deliberative Democracy.* Oxford, UK: Oxford University Press.

Pimentel, D., and Raman, K. (2003). Postharvest food losses to pests in India. In R. Lal, D. Hansen, N. Uphoff and S. Slack (Eds.), *Food Security and Environmental Quality in the Developing World* (pp. 261–268). Boca Raton, FL: CRC Press.

Radjou, N., and Prabhu, J. (2015). *Frugal Innovation: How to do More with Less.* New York: PublicAffairs.

Ritchie, J., Lewis, J., Nicholls, C. M., and Ormston, R. (2013). *Qualitative Research Practice: A Guide for Social Science Students and Researchers.* Los Angeles, CA: Sage Publications.

SAWBO. (2016). Improved method of shea butter processing. Retrieved from http://sawbo-animations.org/video.php?video=//www.youtube.com/embed/7zswmb7oIwM

Shackleton, S., Paumgarten, F., Kassa, H., Husselman, M., and Zida, M. (2011). Opportunities for enhancing poor women's socioeconomic empowerment in the value chains of three African non-timber forest products (NTFPs). *International Forestry Review*, 13(2), 136–151.

Sosa, M. C., Lutz, M. C., Lefort, N. C., and Sanchez, A. (2016). Postharvest losses by complex of *Phytophthora* sp. and *Botrytis cinerea* in long storage pear fruit in the North Patagonia, Argentina. Presented at the III International Symposium on Postharvest Pathology: Using Science to Increase Food Availability, Bari, Italy,

Szulanski, G., Cappetta, R., and Jensen, R. J. (2004). When and how trustworthiness matters: Knowledge transfer and the moderating effect of causal ambiguity. *Organization Science*, 15(5), 600–613.

Tata, J. S., and McNamarra, P. S. (2018). Impact of ICT on agricultural extension services delivery: Evidence from the CRS SMART skills and Farmbook project in Kenya. *Journal of Agricultural Education and Extension*, 24(1), 89–110.

United Nations. (2016). *Transforming Our World: The 2030 Agenda for Sustainable Development.* New York: United Nations.

Van Mele, P. (2011). *Video-Mediated Farmer-to-Farmer Learning for Sustainable Agriculture. A Scoping Study for SDC, SAI Platform and GFRAS.* Ghent, Belgium: Agro-Insight.

Vygotsky, L. (2012). The science of psychology. *Journal of Russian and East European Psychology*, 50(4), 85–106.

Wals, A. E. (2015). Social learning-oriented capacity-building for critical transitions towards sustainability. In R. Jucker and R. Mathar (Eds.), *Schooling for Sustainable Development in Europe* (pp. 87–107). Dordrecht, the Netherlands: Springer.

Walter, A., Finger, R., Huber, R., and Buchmann, N. (2017). Opinion: Smart farming is key to developing sustainable agriculture. *Proceedings of the National Academy of Sciences*, 114(24), 6148–6150.

Wenger, E. (1998). *Communities of Practice: Learning, Meaning, and Identity.* Cambridge, UK: Cambridge University Press.

Wesolowski, A., Eagle, N., Noor, A. M., Snow, R. W., and Buckee, C. O. (2012). Heterogeneous mobile phone ownership and usage patterns in Kenya. *PLoS One*, 7(4), e35319.

13 Development, Demonstration, and Utility of Stored Grain Saving Technologies
A Case Study in India

S. Mohan

CONTENTS

13.1 INTRODUCTION

India's food grain production has witnessed a phenomenal growth since the country's independence 70 years ago. Based on long-term food grain production and the population in different periods, India has been maintaining a gross per capita food grain availability of nearly 197 ± 16 kg throughout the post-independence period. Though grain production has been steadily increasing due to advances in production technology, improper storage results in high losses in grains.

A recent survey conducted by the Indian Council of Agricultural Research (ICAR) concludes that the postharvest losses (PHL) in the food grains range from 6% to 8%, and that of fruits and vegetable range from 12% to 18% (Jha et al. 2015). The total loss is equivalent to INR 920 billion (US$14.2 billion) per year. India loses nearly 12 to 16 million metric tons of food grains annually, worth about US$4 billion. This is sufficient to feed nearly 10% of India's population, meaning 10% of India's food demand can be met by safely storing grains and by minimizing storage losses. The majority of losses occur at farm/home level where nearly 60%–70% of total food grains produced in the country are stored by farmers on their farms. Poor storage facilities and not adopting scientific methods of storing grains are the major cause of such huge losses.

Storage of grains in warehouses is very common in India. The Food Corporation of India (FCI) is the largest food grain trading and distributing agency in India, and probably the largest supply chain management system in Asia. At present, the total food grain storage capacities owned by FCI and organizations hired by FCI is around 94.53 million metric tons (Bhartendu and AnilRaj 2015). Even in these well-maintained warehouses, there is a loss of 0.41% food grain stocks (Girish 1990). Further, phosphine, a widely accepted fumigant, is the most significant means of controlling insect pests in stored grains and processed commodities. The warehouses are solely dependent on phosphine, as no suitable alternatives are available as a tool for management of stored grain pests; indiscriminate and prolonged use of phosphine gas has resulted in the development of inherited resistance in most of the stored grain insects (Sonai Rajan 2015). Though several schemes funded by various agencies over many years stressed "good fumigation" practices to overcome resistance, practical adoption in actual field conditions is still a great question to be answered.

With no alternative to phosphine, there is an urgent need for eco-friendly technologies for stored grain insect management at the farm, home, and warehouse. As food grains cannot be treated with botanicals and kaolinitic clays, there needs to be a way to tackle the insect problems in storage at all levels. Precisely, there is a worldwide need for an insect-free stored grain revolution, which can happen only through innovations, usable eco-friendly technologies that can play a significant role in Integrated Pest Management (IPM), as well as a phosphine resistance management strategy in developing countries.

These situations warrant design and development of useful tools/gadgets for management of insects. Detailed information on such tools developed in India, and efforts made to transfer these tools to the beneficiaries, as well as end results are discussed in this review, which will benefit any countries having a similar climatic situation like India.

13.1.1 DEVELOPMENT OF GADGETS

Loschiavo and Atkinson (1967) developed the concept of exploiting the wandering behavior of many stored product insects via early detection of their presence, and trapping them out of the stored grain in the early period of storage when generally the insect population is extremely low (0–1 number/kg of grain). In India, the author started working in this area from 1992 (Mohan and Gopalan 1992), which led to a series of developments, not only in innovating and developing simple eco-friendly gadgets, but also taking them to the end users across the world.

13.1.2 Technologies Developed

It is well proved that no granaries can be filled with grains without insects, as the harvested produce contains eggs, larvae, and/or pupae because of field carryover infestation, which cannot be avoided, especially in developing countries like India. So, what is required is simple technologies for timely detection of insects in the stored produce and to plan timely control measures.

Many devices have been developed for detection of stored grain insects, some of which are popularly used across the country in households, farms, and warehouses.

13.2 GADGETS DEVELOPED TO DATE (www.mohantrap.com)

A list of key gadgets includes: (1) Tamil Nadu Agricultural University (TNAU) insect probe trap; (2) TNAU pitfall trap; (3) TNAU two-in-one trap for pulse beetle; (4) indicator evice; (5) automatic insect removal bin; (6) ultraviolet (UV) light trap for warehouses; (7) egg removal device (Indian patent no: 198434); (8) stack probe trap (Indian patent no: 284727), and (9) stored grain insect pest management kit.

All these tools can be used for both monitoring and mass trapping of stored grain insects. It is important to note that the presence of even a single live insect in the food grain cannot be tolerated, as they build up and cause an enormous loss in storage due to their high reproductive rate.

13.2.1 TNAU Insect Probe Trap

The use of traps is a relatively new method of detecting and trapping insects in stored grains. The basic components of a TNAU probe trap consist of three important parts: the main tube, the insect trapping tube, and a detachable cone at the bottom. Equally spaced perforations of 2 mm diameter are made in the main tube (Mohan and Gopalan 1992).

  Concept: Insects love "air" and move towards the air. This behavior of the insect is exploited in this technology.

  Method of working: The insect trap has to be kept in the grain (rice, wheat, etc.), vertically with the plastic cone downside. The top cap must be level with the grain. Insects will move towards air in the main tube and enter through the hole. Once the insect enters the hole it falls down into the detachable cone at the bottom. Then there is no way to escape, and the insects are trapped forever. The detachable cone can be unscrewed weekly and the insects can be destroyed.

  Efficiency: TNAU insect traps are excellent insect detection devices in food grains, and more effective in the detection of stored grain insects, namely lesser grain borer *Rhyzopertha dominica* (F.), rice weevil *Sitophilus oryzae* (L.), and red flour beetle *Tribolium castaneum* (Herbst) in stored food grains, both in terms of detection as well as number of insects caught, than the standard normal sampling method (by spear sampling). The detection ratio (trap:normal sample) is higher in the trap than of the normal sampling method by factors ranging from 2:1 to 31:1. The insect catch is also higher in the probe trap than the normal sampling method by factors ranging from 20:1 to 121:1 (Mohan 1993).

They are also good mass trapping devices when used at a density of 2–3 units per 25 kg bin (28 cm diameter and 39 cm length). They should be placed in the top 6 inches (15.25 cm) of the grain, where the insect activity is seen during the early period of storage. They can remove >80% of the insects within 10–20 days (Mohan et al. 2006).

13.2.2 TNAU Pitfall Trap

Pitfall traps are used for capturing insects active on the grain surface and in other layers of the grain as a monitoring and mass trapping tool.

The TNAU model (Mohan et al. 2001) has a perforated lid and cone-shaped bottom, which tapers into a funnel-shaped trapping tube. The commercial model is made of plastic, simple and economical, and easy to handle.

13.2.3   TNAU Two-in-One Model Trap (Mahendran and Mohan 2002)

The probe trap containing the components, namely the perforated tube, pitfall mechanism, a collection tube, and the cone-shaped pitfall trap with a perforated lid and the bottom tapering cone, were combined as a single unit. The combination of probe and pitfall increases the insect-trapping efficiency. It is best suited for pulse beetles, as they are seen typically wandering only on the grain surface. It does not require tedious procedures like coating the inner surface of the pitfall cone with sticky materials before trapping to hold pulse beetles. Beetles are captured alive in this trap, which may facilitate the release of pheromones, thereby attracting more insects (Mohan et al. 2008).

13.2.4   Indicator Device for Timely Detection (Mohan et al. 2001)

The indicator device consists of a cone-shaped perforated cup (3 mm perforation) with a lid at the top. The cup is fixed at the bottom with a container and circular dish, which are to be smeared with a sticky material.

Farmers, before storing their pulses, should take 200 g of pulses to be stored and put them in the cup. When the beetles carried over from the field start emerging, due to their wandering behavior, they enter the perforations, slip off, and fall into the trapping portions. As they stick to the sticky materials, farmers can easily locate the beetles and can take out the bulk-stored pulses for sun drying. A device with 2 mm perforations can be used for cereals.

13.2.5   TNAU Automatic Insect Removal Bin (Mohan 2000)

The TNAU insect removal bin can remove insects automatically. The structure has four major parts, namely an outer container, an inner perforated container, a collection vessel, and a lid. The model exploits wandering behavior of stored product insects, as well as the movement of these insects towards well-aerated regions. The grains are held in the specially designed inner perforated container. The space between inner and outer container provides good aeration for the insects. Insects, while wandering, enter the perforation to reach the aerated part and, while doing so, slip off and fall into the collection vessel through a pitfall mechanism provided in the collection vessel. In order to quickly collect the insects, as and when they emerge from grains, perforated (2 mm) rods are fixed in the inner container.

The container will be useful for storing rice, wheat, broken pulses, coriander, etc. Insects such as rice weevil, lesser grain borer, red flour beetle, and saw-toothed beetle, which are commonly found attacking stored grains, can be removed automatically by storing grains in this container. Within a very short period of 10 days, a majority of the insects (more than 90%) can be removed from the grains. The containers are available in 2 kg, 5 kg, 25 kg, 100 kg, and 500 kg capacities.

Efficiency: Grains (paddy and sorghum) stored in the automatic insect removal bin (100 kg and 500 kg) recorded only 1%–4% damage by insects, compared to 33%–65% damage in the ordinary bin after 10 months of storage. The population of insects (*R. dominica* and *S. oryzae*) ranged from 0–2/kg in grain stored in a 100 kg automatic insect removal bin, compared to 5–191/kg in the ordinary bin after 10 months of storage (Mohan 1994).

13.2.6   UV – A Light Trap for Grain Storage Warehouses

The UV light trap mainly consists of an UV source (a 4 W germicidal lamp). The lamp produces UV rays of peak emission around 250 nm. The UV light trap can be placed in a food grain storage warehouse at 1.5 m above ground level, preferably in places around warehouse corners, as it has been observed that the insects tend to move towards these places during the evening hours. The trap can be operated during the nighttime. The light trap attracts stored product insects of the paddy, like, *R. dominica* , *T. castaneum*, and cigarette beetle *Lasioderma serricorne* in large numbers. Psocids, which are a great nuisance in the warehouse, are also attracted in large numbers. Normally 2 UV light traps per 60 m × 20 m (length × breadth) warehouse at a height of 5 m is suggested. The trap is ideal for use in warehouses meant for long-term storage of grains, whenever infested stocks arrive in warehouses, and during post-fumigation periods to trap the resistant strains and leftover insects to prevent the buildup of pest populations. In warehouses of frequent transactions, the trap can be used for monitoring (Mohan and Rajesh 2016).

Efficiency: It has been found that two traps kept at the corners of a 60 m × 20 m × 5 m warehouse can catch around 200 insects/day, even in a warehouse where normal sampling did not show any insect presence, thus indicating its effectiveness as a monitoring and mass trapping device. It has been recorded that around 3000 *R. dominica* adult insects could be trapped on a single day by a single trap kept in a paddy warehouse (Mohan et al. 1994).

13.2.7 A DEVICE TO REMOVE INSECT EGGS FROM STORED PULSE SEEDS (INDIAN PATENT NO. 198434)

Pulses are more difficult to store than cereals, as these suffer a great amount of damage during storage by pulse beetle *Callosobruchus* spp. The main source of infestation by pulse beetle is its carry-over damage from the field to stores, which is well known. The present invention is a prototype of a gadget that can successfully crush the eggs of the pulse beetle species *C. chinensis* and *C. maculatus*, which attack stored pulses. The gadget has an outer container and an inner perforated container with a rotating rod fixed with plastic brushes on all sides (Hategekimana and Mohan 2013).

Due to the slashing action of the brushes in the rotating rod, the eggs get crushed and thus the damage is prevented. The treatment does not affect the germination of seeds. The device is also useful for rice, broken pulses, sorghum, and maize to remove adult insects as well as crush free-living larvae, pupae, and eggs laid externally (Figure 13.1).

13.2.8 TNAU-STACK PROBE TRAP FOR MONITORING STORED PRODUCT INSECTS IN THE WAREHOUSE (INDIAN PATENT NO. 284727)

This invention relates to a device for detecting stored grain insects in bag stacks, which comprises a main hollow tube having a diameter in the range of 1.8 cm to 2.0 cm, with equally spaced

FIGURE 13.1   TNAU insect egg remover.

FIGURE 13.2   TNAU stack probe trap.

perforations in the range of 1.8 mm to 2 mm on its upper portion. There is a bend at one end, which ends in a transparent collection unit to collect the insects falling down from the bend, the other end of the main tube being closed (Hategekimana et al. 2013) (Figure 13.2).

Advantages of the invention include:

1. The device is useful in detecting stored grain insects in bag stacks in food grain warehouses, without any damages to sacks.
2. The device does not require any bait materials to trap insects.
3. The device is useful in studying the distribution pattern of stored-product insects in various layers of bag stacks.
4. The device will be useful to validate the effect of fumigation by using it immediately after fumigation, in different layers of the fumigated stacks.

13.2.9   TNAU STORED GRAIN INSECT PEST MANAGEMENT KIT

The author has developed an extension kit named a "TNAU Stored Grain Insect Pest Management Kit," containing prototypes of all the devices, along with a CD-ROM giving details about the devices and how to use them. This kit will be of great use in the popularization of the technologies across the country. The kit will be an ideal "hands-on training" tool for education, extension centers, save grain centers, and for private/public warehousing. This TNAU kit is the first of its kind in the world (Mohan 2007).

13.3   POPULARIZATION OF THE DEVICES

13.3.1   STEP 1: THROUGH COMMERCIALIZATION OF TECHNOLOGY

The author's innovations led to the birth and the growth of four entrepreneurs in India to manufacture and market the gadgets:

- KSNM Marketing (2002): SF No.29/1B, Ona Palayam, Siruvani Water Line Road, Dheenam Palayam Post, Coimbatore – 641 109. Tamil Nadu, India. Web: www.ksnmmarketing.com. Email: ksnmmarketing@hotmail.com
- Melwin Engineering (2011): 18/2, Gandhi Street, Bharathi Nagar, Podanur (PO), Coimbatore – 641 023. Tamil Nadu, India. Email: anitathomascs10@gmail.com
- Khusboo Enterprises (2014): AZIZ Complex, Panbazar, Guwahati, Assam, India. Email: ravi_agarwal8@yahoo.com
- Bhuvi Care (P) Ltd. (2014): Sipcot Industrial Growth Centre, Gangaikondan, Tirunelveli – 627 352. Tamil Nadu, India. Email: bcpl2002@gmail.com

13.3.2   STEP 2: THROUGH PUBLICATIONS

1. Research Articles: 108
2. Popular Articles: 120

FIGURE 13.3   TNAU trap in public exhibition.

3. Technical Bulletin/Booklets published:
   a. English: 16
   b. Tamil (Local language): 8
4. Books written: 6
5. Patent obtained:
   a. Device to remove insect eggs from stored pulse seeds (Indian patent no. 198434)
   b. Trap for monitoring stored product insects in the warehouse (Indian patent no. 284727)

13.3.3   STEP 3: EXTENSION VIA WEBSITES

1. www.mohantrap.com
2. http://agritech.tnau.ac.in/crop_protection/crop_prot_crop_insect_sto_trap.html

13.3.4   STEP 4: THROUGH PRACTICAL DEMONSTRATIONS, EXHIBITIONS, AND CASE STUDIES

Several demonstrations and exhibitions have been conducted (>500) with case studies (Figure 13.3).

13.4   A CASE STUDY WITH TNAU PROBE TRAP

Detailed systematic feedback studies on TNAU probe trap were conducted by the author in Pallapalayam Village, Palladam in Coimbatore, Tamil Nadu, India with Canadian International Development Agency (CIDA) during 2005.

A detailed investigation was made on the effect of TNAU probe trap for management of insect pests in stored rice in Pallapalayam Village. The various phases of the study program included:

1. Baseline survey of the village
2. Awareness program presented to the agricultural officers of the target village
3. Awareness program presented to the farmers
4. Feedback studies at farmers' holdings in target village

13.4.1   BASELINE SURVEY OF THE VILLAGE

A baseline survey on the storage practices followed by the households/farmers has been carried out in the Pallapalayam village.

13.4.1.1 Survey Report

A total of 30 households were surveyed in the village. Personal profiles of the respondents were collected.

The respondents are from all three age groups, that is, young, middle and old (Table 13.1). This complexity will help in dissemination of technology to all age groups. The TNAU gadget technologies are simple to understand by any person.

About 87% of the respondents are literate, with the majority of this group found to have high school education (Table 13.2). This condition will be more favorable for the dissemination of newer technologies in the adopted village.

Among the respondents surveyed, there is an equal proportion of agriculture and non-agriculture groups (50% each; Table 13.3). As food grains are stored both at farms and at home, this proportion will be a great incentive for program implementation, as programs for nutritional security should start at the individual level, whether it is at the farm or at home.

Surveys on the storage methods adopted by the households revealed that bag, plastic drum, and metal drum are the common storage structures used by the respondents (Table 13.4). The percentage use of the above structures was 62%, 23%, and 6% for the bag, plastic drum, and metal drum, respectively. Six percent of the respondents use both bag and drum. The average storage period was three months. This period is optimum for the study proposed, as the insects can multiply within 40 days and can cause an enormous loss at the end of 3–4 months if unnoticed and if proper management methods are not adopted.

TABLE 13.1
Age Group of the Participants

Variable	Category	No (n = 30)	Percentage (%)
Age	a. Young (>35 yrs)	4	13.00
	b. Middle (35–45)	8	27.00
	c. Old (>45)	18	60.00

TABLE 13.2
Education Level of the Participants

Variable	Category	No (n = 30)	Percentage (%)
Education level	a. Illiterate	4	13.00
	b. Literate		
	Primary	1	3.00
	Elementary	5	17.00
	Middle	7	23.00
	High school	8	27.00
	Higher secondary	3	10.00
	Graduation	2	7.00

TABLE 13.3
Primary Occupation of the Respondents

Variable	Category	No. (n = 30)	Percentage (%)
Occupation	a. Agriculture	15	50.00
	b. Non-agriculture	15	50.00

TABLE 13.4
Storage Practices—Types of Storage

Category Storage	No. (n = 30)	Percentage (%)	Average Value of Storage Structure (Rs)#	Average Storage Capacity (Kg)	Average Storage Period (months)
a. Bag	19	63	25	75	3.14
b. Plastic drum	7	23	220	30	2.88
c. Metal drum	2	6	125	75	3.00
d. Bag and drum	2	6	Bag - 25	75	3.00
			Drum 125	30	2.00

Rice is the major food grain stored by all respondents.

Rice is the predominant food grain stored by all respondents. Only a few (<1%) stored split pulses and maize, that too for short-term use (1 month). As rice is attacked by several insects during storage and, further, rice cannot be sun-dried as it will affect the cooking quality, the village will be ideal for the adoption of TNAU gadgets, as many of the gadgets remove insects from grain automatically.

A study on the adoption rate by the respondents regarding various recommended control practices (Table 13.5), showed that 50% of the respondents adopted the method of periodical cleaning, followed by shade drying.Only one among the 30 respondents uses the TNAU gadgets (i.e., TNAU probe trap). So popularization through village level meetings and assisting selected people in the village to adopt TNAU gadgets will have a very good impact in this village.

Among the 30 respondents surveyed, only four persons are aware of TNAU gadgets (Table 13.6), and those persons were aware of only one trap (i.e., TNAU probe trap). The source through which

TABLE 13.5
Adoption of Recommended Practices for Postharvest Insect Management

	Adoption (n = 30)			
	Yes		No	
Practices	Number	Percentage	Number	Percentage
Cleaning	15	50	15	50
Periodic drying				
Chemical treatment (fumigation with phosphine)	—	—	30	100
Plant products with neem			30	100
Gunny bag impregnation with insecticides	—	—	30	100
Pesticide spray over bags	—	—	30	100
TNAU gadgets	1	3.0	29	97.00

TABLE 13.6
Awareness about TNAU Gadgets

	Awareness (n = 30)			
	Aware		Not Aware	
Category	Number	Percentage	Number	Percentage
TNAU gadgets	4	13	26	87

they had come to know the trap was farm/urban meetings organized by TNAU, as well as by the State Department of Agriculture. Out of this four, only one uses the technology of TNAU probe trap for timely detection. Timely detection is followed by timely cleaning and shade drying, which is the best method of stored product insect management in rice storage. Hence, as the majority of the respondents of the target village are not aware of the TNAU technologies, this village was selected for further feedback studies with TNAU probe trap.

13.4.1.2   Feedback Study on TNAU Probe Trap

13.4.1.2.1   Stored Product Insect Loss Estimation in the Village

The loss caused by insects to rice in terms of damage and live insect presence was estimated by a random survey. For quantitative damage estimation, 100 rice grains were randomly taken from each household and the number of weevilled grains were counted and classified as below:

Category	Percent Damage (Range)
Low	1–4
Medium	5–9
High	>10

For assessing the insect population, TNAU probe traps were used (Figure 13.4). The traps were placed in all households, and one week later they were removed and observed for insects and classified as below:

Category	Population/Trap
Low	1–4
Medium	5–10
High	>10

FIGURE 13.4   Feedback studies on TNAU probe trap in households.

TABLE 13.7

The Results of the Feedback Study on TNAU Probe Trap: (Sample Size = 30)

	Nil		Low		Medium		High	
Storage Loss	No	%	No	%	No	%	No	%
a. By sampling	11	37	15	50	3	10.0	1	3.0
b. By trap	16	54	13	43	—	—	1	3.0

It is clear that there is a good amount of insect infestation in the rice stored in the village. About 63% of the samples had insect infestation and 46% had live insects. This indicates there exists the good potential for the use of TNAU gadgets to reduce storage losses in the target village (Table 13.7).

13.4.1.2.2   Insect Species Present

The survey conducted to find out an overall view of the presence of insects in the stored rice in the village showed interesting results.

Respondents who had incidence in their stock (Sample Size = 30)

Insect	Number	Percentage
T. castaneum	7	23
O. surinamensis	12	40

It is clear that the random survey conducted using traps revealed that almost 63% of respondents who have rice storage (in bag or drum) had live insects in the product. Two species, namely *T. castaneum* and *Oryzhaephillus surinamensis*, occurred in rice storage. It is clear that there is insect multiplication in the rice stored by the respondents. Hence there is a good scope to demonstrate and prove the effectiveness of TNAU gadgets in this village.

13.4.1.2.3   Problems Faced by the Respondents During Storage of Grains

The survey on the problems based on the farmers gave a clear picture of important aspects (i.e., they were not able to identify the presence of insects in their stored grain, and only when there is severe damage they can note the presence of insects). Among the 30 respondents, 28 (94%) posed this problem to the scientist.

This clearly indicates the fact that early detection of insects in grain storage is important and as TNAU has good detection and mass trapping technologies for varied storage conditions/methods, the village Pallapalayam is an ideal village to create awareness, as well as to make people adopt TNAU gadgets. Hence awareness programs and feedback studies were initiated in this village.

13.4.2   Awareness Program for the Agricultural Officers of the Target Village

An awareness program on the use of various TNAU gadgets for stored grain insect control was conducted to the Agricultural Officers and Assistant Agricultural Officers of the State Agricultural Department, Palladam Taluk, Coimbatore District, Tamil Nadu, in which the target village is located. About 25 participants attended the meeting, and all the gadgets were demonstrated to them. The TNAU kit box for stored grain insect management was also demonstrated.

FIGURE 13.5   Awareness program for the farmers.

13.4.3   Awareness Program for the Farmers

An awareness program for the farmers of the village was held on December 22, 2005 in the presence of Dr. G.S.V. Raghavan, Project Director, CIDA. Around 60 participants, mostly women, attended the program (Figure 13.5).

13.4.4   Feedback Studies at the Farmers' Holdings in the Target Village

Rice can be shade dried, or can be cleaned periodically (once in a month) to avoid insect infestation. In this situation, use of traps will be a boost to the households. A study was conducted using TNAU probe trap to find its effectiveness in reducing the losses of rice grains caused by stored grain insects in bins (25 kg–30 kg capacity) in the target village.

Fifteen families were selected for this study in the target village, based on the information obtained through the baseline survey. The commodity stored was rice and the storage structure was bins (tins, metal wooden bins, etc.).

Two traps were placed in each bin, covering the top 6 inches (~15 cm) of the grain layer, where the activity of insects is seen during the initial period of storage. The traps were placed immediately after storage of rice.

Traps were periodically cleaned of the insects (once a week) and replaced again. The feedback study period was three months, as generally people store the produce only for three months in the target village.

The number of insects trapped was recorded and species identified. At the end of the study period, 1 kg of random samples of rice grains was taken from the bin and sieved to assess the insect population. From this sample, 1000 grains were observed for insect damage.

Parallel to this, insect population and loss estimate in ordinary bins (without a trap) were also assessed by randomly selecting 15 households in the target village for comparative feedback analysis.

13.4.4.1   Significant Outcome

1. The major insect species recorded were *S. oryzae* and *Oryzaephilus* spp.
2. There was very good reduction in population of the insects, as well reduced rice grain loss in the bins where TNAU probe traps were used, compared to bins without a trap (Table 13.8).

TABLE 13.8

Insect Population and Percent Grain Loss in Bins with TNAU Probe Traps and in Control (Bins Without a Trap) in Pallapalayam Village

Sl. No.	Particulars	Insect Population (No./kg)	Grain Loss (%)
		(Mean of 15 Observations)	
1.	Ordinary bin + probe trap	4.73	2.65
2.	Ordinary bin alone (control)	7.25	4.58

3. Cost-benefit ratio:

A loss of around 2% was prevented. It is good enough for safe storage of grain in a single household. Loss generally means the presence of weevilled/broken grains, which are the primary causes of further insect multiplication. Further, there is the prevention of quality deterioration. Taking into consideration the current price of one TNAU probe trap (INR 100) and the selling price of clean whole good quality rice grain (INR 60), the benefit is quickly substantial, in just three months' time. Generally, in the target village, the storage of rice grain is around 25 kg–50 kg per family, which can be used for a period of 3–5 months. An additional benefit is that life of the trap is approximately five years, since it is made of food grade quality stainless steel material. As saving grain should start from every household, as a grain saved is a grain produced, the feedback assures significant in Post Harvest Technology (PHT) science.

13.5   IMPACT OF TRANSFER OF TECHNOLOGY

Technology 1: TNAU probe trap:
- Around 0.3 million people in India use TNAU insect trap
- Introduction of TNAU trap in Africa (Ethiopia, Rwanda, Nigeria, and Egypt), Turkey, Thailand, and France (Amsalu Debebe et al. 2008)

Technology 2: TNAU automatic insect removal bin:
- Around 5,000 farmers in the northeastern zone of India use the insect removal bin for paddy seed storage
- This is the first time a technology has gone to the northeastern zone of India, which is the current priority area for agricultural development in the country

Technology 3: TNAU UV light trap:

Around 1,000 UV light trap units are currently in use in India by:
   a. Food Corporation of India
   b. Madaus Pharmaceuticals Pvt. Ltd., Goa
   c. Saraf Trading Corporation Pvt. Ltd., Cochin
   d. SKM Siddha and Ayurvedic, Tamil Nadu
   e. Cadburys India Ltd., Dharapuram, Tamil Nadu
   f. Jayanthi India Spices Ltd., Coimbatore
   g. Mahyco Maharashtra Hybrid Seeds Company Ltd.
   h. Ulavan (Farmers) Producers company, Erode, Tamil Nadu

Technology 4: TNAU stack probe trap (Indian patent no. 284727):

Around 500 units are currently in use in India by:
   a. Ulavan (Farmers) Producers Company, Erode, Tamil Nadu
   b. Gubba Cold Storage Ltd., Hyderabad
   c. Jayanthi India Spices Ltd., Coimbatore

Technology 5: TNAU insect egg remover (Indian patent no: 198434):
> Number of units currently in use include:
> > a. Machine operated: 3 (mainly rice traders/merchants buy this machine)
> > b. Hand operated: 2 (Self Help Groups (SHGs))

Technology 6: TNAU kit:
- Around 300 agricultural colleges/farmer'training centers across the country are using the TNAU Kit for teaching and training
- Around 400 schools enrolling 12,000 students in Tamil Nadu are using TNAU kit for teaching school children

13.5.1 IMPACT ON SCHOOL EDUCATION

Besides being useful to the farmers, households, and warehouse managers, TNAU insect trap technologies paved the way for improving the scientific temper of school children, made possible only by extensive popularization activity done by the author:

1. Seventh-standard student of Chinmaya Vidyalaya, Trichy, Tamil Nadu: Mr. M. Abhilash has won two gold medals in the INTEL–IRIS competition and a World Intellectual Property Organization (WIPO) Award for young inventors by making TNAU traps using waste materials (Bisleri bottles).
2. Students of the Senthil Matriculation School in Dharmapuri, Tamil Nadu won a medal at the 10th National Children's Science Congress in 2002 by developing their own model of insect trap from the inspiration they got on seeing the probe trap model.
3. U.R.C. Palaniammal Matric Higher Secondary School, Sengodampalayam, Erode, Tamil Nadu has won a prize and INR 1000 cash award for the model "Protection of Food Grains" in exhibition cum competition "Genesis 2003" organized by P.K.R. Arts College for Women, Gobichettipalayam, Tamil Nadu.
4. Mr. R. Vignesh of Padma Sheshadri Balamandir Higher Secondary School, Chennai, Tamil Nadu won the best project award for making TNAU traps in plastic waste materials during "School Annual Meet 2009."

Impressed by the creative activity of the school students of Tamil Nadu, based on the technologies developed, the author was appointed as the chairman of the committee to revamp and revitalize agricultural school education in Tamil Nadu. The old curriculum structure, which was in practice in Tamil Nadu for around the last 20 years, was revamped and a common textbook series, Agricultural Practices I and Agricultural Practices II (for 11th and 12th standard schools) was introduced in 2010–2011 and 2011–2012, respectively. For the first time in the curriculum, a practical guide was introduced for 11th and 12th standards in Tamil Nadu. The 11th Agricultural Practices I, which is currently read by approximately 12,000 school students enrolled in around 400 schools in Tamil Nadu, has a chapter on the "Importance of Postharvest Technology" in which the TNAU probe trap is highlighted.

13.5.2 IMPACT ON POSTGRADUATE EDUCATION

A simple and rapid technique named cup bioassay, to determine if natural products are repellent or attractive to stored product insects, was developed during the training undergone by the author in the Cereal Research Center, Canada. This has been included in the postgraduate curriculum in Tamil Nadu (Mohan and Fields 2002).

13.6 FUTURE THRUSTS AND CONCLUSION

It is very clear from the above outcome of the popularization activities carried out to date that there is a very good response to the gadgets from diverse end users (home/farm/warehouse). If one question may arise, it might be why is it that only very few (0.3 million) have benefitted? This is a great question, for which the author's answer is that there is a need for developing entrepreneurs across India, as currently only four entrepreneurs have taken up the gadgets, and among them only two are active. The author is tirelessly working to attract entrepreneurs, since at least 2–3 in each region of India are required to easily make the technologies to reach around 300 million users in a very short time.

Efforts are ongoing, and it will happen one day as the TNAU gadget technologies have proved beyond doubt that they form an integral part of Insecticide Resistance Management (phosphine resistance) as well as IPM in grain storage.

A pair of insects, through three generations per year, has the biotic potential to produce 1,500,000 offspring, which will consume 1,500,000 kernels of rice (amounting to 30 kg of rice), in addition to causing quality deterioration and seed viability loss (36%–41%). The above-described gadgets developed by TNAU no doubt will play a significant role in saving grain from insects at all levels of storage (home/farm/warehouse), not only in India but also in all developing countries as tools of management and tools of education for postharvest science.

REFERENCES

Debebe, A., S. Mohan, S. Kuttalam and R. Nagappan. 2008. Cross border technology transfer a case study with TNAU trapping technologies in Gondar, Ethiopia, Africa for the management of stored product insects. *Research and Reviews in Biosciences*, 2(1): 48–50.

Bhartendu, K.R.C. and T.A. AnilRaj. 2015. Agricultural storage infrastructure in India—An overview. IOSR. *Journal of Business and Management*, 17(5): 37–43.

Girish, G.K. 1990. Post harvest grain technology in India. In: *Proceedings of Regional Workshop on Warehouse Management of Stored Grains* (Eds.), G.K. Girish and Ashok Kumar. pp. 145–160. Ministry of Food and Civil Supplies, Government of India, New Delhi, India.

Hategekimana, A. and S. Mohan. 2013. Determining the effectiveness of insect egg remover in removal of adult insects in stored grains. *Asian Academic Research Journal of Multi-disciplinary*, 1(10): 266–277.

Hategekimana, A., S. Mohan, K. Ramaraju and V. Thirupathi. 2013. Efficiency of stack prole trap for detection, validation of fumigation and distribution of stored grain insects in Bag stacks. *Journal of Renewable Agriculture*, 1(3): 39–43.

Jha, S.N., R.K. Vishwakarma, T. Ahmed, A. Rai and A.F. Dixit. 2015. Report on assessment of quantitative harvest and post harvest losses of major crops and commodities in India. ICAR- All India Co-ordinated Research project on Post harvest technology, ICAR-CIPHET, Ludhiana, India.

Loschiavo, S.R. and J.M. Atkinson. 1967. A trap for detection and recovery of insects in stored grain. *Canadian Entomologist*, 99: 1160–1163.

Mahendran, K. and S. Mohan. 2002. Technology adoption, estimation of loss and farmers behaviour in pulse storage. A study in western Tamil Nadu. *Pestology*, XXVI(11): 35–38.

Mohan, S. 1993. Studies on the detection and management of stored product insect pests of rice. PhD thesis, Tamil Nadu Agricultural University, Coimbatore, India.

Mohan, S. 1994. A storage container model for automatic removal of stored product insects from grains. *Bulletin of Grain Technology*, 32(2): 178–181.

Mohan, S. 2000. Insect removal bin could help farmers in developing countries. *Resource*, 7(9): 6.

Mohan, S. 2007. Ecofriendly postharvest technologies for management of stored grain insects. *Green Farming*, 1(2): 45–47.

Mohan, S., C.T. Devadass and D. Mahendran. 2001. New devices for pulse beetle management. *Plant Protection Bulletin, Government of India*, 53.3(4): 27–32.

Mohan, S. and P. Fields. 2002. A simple technique to assess compounds that are repellant (or) attractive to stored product insects. *Journal of Stored Product Research*, 38: 23–31.

Mohan, S. and M. Gopalan. 1992. Stored grain insect trap. *Rice India*, 2(8): 15–16.

Mohan, S., M. Gopalan, P.C. SundaraBabu and V.V. Sreenarayanan. (1994). Practical studies on the use of light trap and bait trap for management of *Rhyzopertha dominica* in rice warehouses. *International Journal of Pest Management*, 40(2): 148–152.

Mohan, S. and A. Rajesh. 2016. Use of light traps in a phosphine resistance management strategy for *Tribolium castaneum* (herbst) in Indian grain storage warehouses. In: *Proceedings of The 10th International Conference on Controlled Atmosphere and Fumigation in Stored Products*, New Delhi, India, November 7–11, Ref. No.70-4.

Mohan, S., S. Sivakumar, Z. Kavitharaghavan, S. Venkatesh and G.S.V. Ragavar. (2008). A new trap model to increase the trapping of *Cryptolestes ferrugineus* (Coleoptera: Laemophloeidae) in wheat filled containers. *Madras Agricultural Journal*, 95(7–12): 390–393.

Mohan, S., S.S. Sivakumar, S.R. Venkatesh and G.S.V. Ragavan. (2006). Simple technique to increase the sensitivity of probe traps in detecting *Cryptolestes ferrugineus* in stored wheat. *Candian Entomologists*, 138(5): 733–735.

Sonai Rajan, T. (2015). Studies on frequency, distribution and molecular analysis of phosphine resistance, field ecology and population genetics of *Tribolium castaneum* (Herbst) and *Sitophilus oryzae* (L.) in three major southern states of India. PhD thesis, Tamil Nadu Agricultural University, Coimbatore, India, p. 258.

Section IV

Postharvest Extension Models

14 ADMI Village
Research in Action

Alex Winter-Nelson, Mindy Spencer,
Sarah Schwartz, and Ashley Nagele

CONTENTS

14.1 INTRODUCTION

Walking down the main road in the village of Dih Sarsauna is like walking through any one of thousands of communities in rural India. The road is dirt. The houses that line the street are made of wood or brick, many with thatched roofs. The majority of homes lack running water or indoor plumbing. Impossibly overloaded water buffalo, donkeys, wagons, and motorcycles crowd the street. Harvested crops line the side of the road, drying in the sun or waiting for a buyer.

This is not an unusual village in the state of Bihar, where 80% of the population depends on agriculture, and many families earn less than US$1.25 a day per person. Bihar is simultaneously one of the poorest and the most populous states in India.[*] Farm sizes are small, typically one hectare or less, and producers use a combination of simple tools, manual labor, and traditional methods to

---

[*] Majority of the population in Bihar is rural with annual per capita income of USD 250

FIGURE 14.1 Woman winnowing grain with fan.

harvest, dry, and process their crops (Ranganathan 2014).* Walking down the paths of Dih Sarsauna, it's not unusual to see grain sun-drying on plastic sheets, sometimes aided by household electric fans (Figure 14.1).

While much has been made of the Green Revolution that transformed Indian agriculture, especially in the states of Punjab and Haryana, this increase in crop productivity bypassed Bihar and the areas that immediately surround it in India, and in the neighboring countries of Bangladesh and Nepal. Whatever Bihari farmers manage to harvest, they are hard-pressed to market for a profit. Bihar does not have the necessary infrastructure of roads to support efficient transport of goods or crops. At the same time, farms cannot delay sales after harvest, as crop storage and processing facilities are not accessible to most people (Figures 14.2 and 14.3).

Nationwide, crop losses in India from improper postharvest handling are estimated to be as high as 35%. In Bihar, postharvest losses (PHL) of cereal grains are estimated at US$676 million (IL&FS 2014). PHL is a problem that many farmers just accept as a cost of doing business. Indeed, many farmers are unaware of the negative impacts storage methods can have on the quantity and quality of their food; however, by introducing affordable and scalable technologies, PHL becomes a solvable problem in rural India. The scale of the loss means low cost solutions can have a dramatic impact. The demonstration village at Dih Sarsauna was established to demonstrate how changes in postharvest systems can change people's lives by preserving more of the crops they grow (Figure 14.4).

---

* 83% of farmers having less than 1 hectare. 90% of the cropped land is used for food grains (rice, wheat, maize)

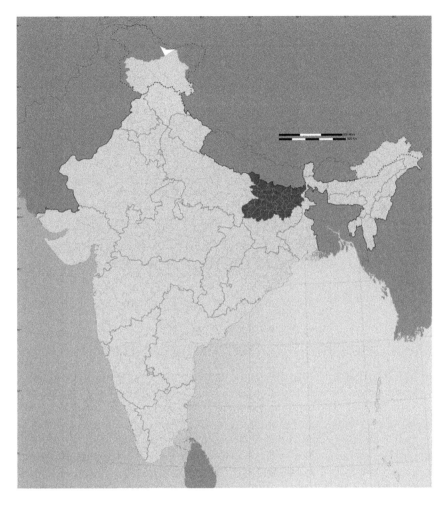

FIGURE 14.2   Location of Bihar in eastern India.

FIGURE 14.3   Three women with thresher.

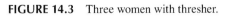

FIGURE 14.4　Sun-drying grain.

14.2　DRAWING ATTENTION TO A WORLDWIDE PROBLEM

In January 2011, the Archer Daniels Midland Company (ADM) provided a generous gift to the University of Illinois at Urbana-Champaign to establish the ADM Institute for the Prevention of Postharvest Loss (ADMI). The institute was created in response to the rising amounts of staple crops lost each year in food chains around the world. The institute serves as an international information and technology hub for evaluating, creating, and disseminating economically viable technologies, practices, and systems that reduce PHL of commodities, including rice, wheat, maize, and oilseeds. Putting research into action is at the core of ADMI's mission and priorities. The institute has conducted applied research on postharvest issues in areas of critical need in the developing world. Partnerships with universities and research institutes around the world create a foundation for ADMI research, outreach, and capacity building to achieve impact. That network of partners is well-represented in Bihar, where the need for PHL prevention is manifest.

14.3　THE NEED FOR POSTHARVEST LOSS PREVENTION IN INDIA

According to the Food and Agriculture Organization of the United Nations (FAO), about one-third of the food produced for human consumption globally is lost or wasted, amounting to approximately 1.3 billion tons per year (FAO, 2011). In South Asia, the FAO estimates about 20% of cereal grains are lost, with 40% of those losses occurring postharvest (Agrilinks 2018).

The Green Revolution transformed India's agriculture sector in the 20th century, allowing the country to grow enough food grain to feed its population and become a leader in global trade of commodities. Despite the success of the first Green Revolution, poverty and food insecurity persist for people in many parts of the country. National and international leaders have called for a "second Green Revolution" as a gridlock of challenges continues to dampen the success of India's rural economy. Indian Prime Minister Narendra Modi has launched an initiative to double Indian farmers' income by the year 2022. One way to work toward this goal is to preserve the crops India's farmers already produce.

While farmers in some parts of India have access to modern grain processing, storage, and transportation, there are stark differences across India's states. According to a 2013 study on farmer income in India submitted to the Indian Ministry of Agriculture, farm household annual income in Bihar is around INR 44,172 (US$663). In Punjab, farm households earn about INR 217,450 (US$3,264) annually (Ranganathan 2014).

14.4   ESTABLISHING THE ADMI VILLAGE

ADMI researchers began studies of PHL in Bihar in 2014 in collaboration with universities in the state. In August 2015, ADMI took a step to move academic research into practical application with the establishment of the ADMI Village in Dih Sarsauna, a village of 2,800 people, many of whom rely on farming a variety of crops, including cereal grains, fruits, and vegetables. The ADMI Village project was officially launched in Patna, Bihar, by Dr. Robert Easter, President Emeritus of the University of Illinois and Chairman of the ADM Institute External Advisory Board; Shri Radha Mohan Singh, Honorable Agriculture Minister of India; and other senior officials from the Indian government and the Indian Council of Agricultural Research.

ADMI mobilized teams of experts from Bihar Agricultural University (BAU) in Bhagalpur, Dr. Rajendra Prasad Central Agricultural University (DRPCAU) in Pusa, and the Borlaug Institute for South Asia (BISA)—Pusa for research and outreach activities in Bihar, concentrated around the ADMI Village. While the research arm of the Bihar work is led from the University of Illinois at Urbana-Champaign, local partners guide and manage village operations, including farmer trainings, public outreach, and field testing of new technologies. Reaching out from the demonstration village, ADMI, together with BAU, DRPCAU, and BISA, conducts research and outreach in more than 80 villages across five districts in Bihar to raise awareness about PHL and provide information for improved grain management practices.

ADMI researchers working with BAU, DRPCAU, and BISA conducted a baseline survey of 800 households in Bihar. This initial baseline research confirmed that while villagers were aware of common PHL prevention techniques, including mechanized threshing, grain drying, and milling, most had not used them. Farmers were asked to self-report their storage loss levels, which most estimated at between 1% and 2%. However, it was clear to the researchers that farmers were not including all losses from mold or moisture. Indeed, later studies would reveal that large shares of maize harvests were compromised by mold, but farmers did not recognize this as a loss, even though it creates a health hazard and diminishes the sale price of their grain. Most farmers in the sample were storing grain for home consumption—farmers there grow about 35% of the grain they consume—so even a small reduction in loss could make a real difference in household food levels (Baylis et al. 2016).

14.5   ADMI GRAIN HANDLING SYSTEM: POSTHARVEST TECHNOLOGY IN THE ADMI VILLAGE

A goal for the ADMI Village project was to release a complete PHL technology solution that is practical, affordable, and scalable. Effective PHL technology needs to provide a mechanism for both drying and storage. Coupling adequate drying with reliable storage is essential to preserving both quality and quantity of grain, ultimately leading to higher premiums at the point of sale. Quality losses resulting from poor postharvest management cause farmers to lose nearly 30% of the price premium they could have earned from their grains (YieldWise 2018). ADMI's research into various solutions for postharvest drying and storage resulted in the introduction of the ADMI Grain Handling System to the ADMI Village.

14.5.1   ADMI GRAIN HANDLING SYSTEM COMPONENT 1: STR DRYERS

An essential element of this project was finding and adapting a small-scale dryer for use in local conditions. ADMI selected the STR dryer, a low cost, Vietnamese-designed dryer that has a half-ton capacity and can decrease moisture levels from 20% to 12% in four hours, as a high-impact, scalable solution for PHL prevention. This dryer was initially identified through ADMI's cooperation with another institution, Bangladesh Agricultural University.

14.5.2 ADMI Grain Storage System Component 2: Hermetic Grain Storage Bags

Poor cereal grain storage can result in losses as high as 50% to 60%. Throughout South Asia, and in much of the developing world, food grains are sun-dried and then stored in locally produced jute bags. Jute storage bags leave grain highly susceptible to moisture, rodents, and insects—the leading causes of PHL. Elevated moisture content can lead to the growth of molds and mycotoxins, which result in both quantitative and qualitative losses. Health effects that have been linked to mycotoxins range from acute toxicosis and death to stunting in children and multiple system (gastrointestinal, neurological, immunologic) damage from chronic exposure. Sufficient drying and proper storage are key to mycotoxin prevention (Kumar and Kalita 2017). If grain is properly dried and then stored in hermetically sealed bags, moisture content remains sufficiently low to prevent emergence of mycotoxins. In addition to sealing the bags against changes in humidity, the bags prevent rodents from smelling and destroying the grain and stops insects from infesting the grain.

14.5.3 ADMI Grain Storage System Results

ADMI researchers have studied the impact of combined dryers and hermetic storage bags in laboratories and in farmers' homes. Research conducted by the ADMI Bihar research team demonstrated that farmers experienced a 14% reduction in storage losses when using hermetic bags in combination with STR dryers (ADMI 2017). The system is effective and economical (Figure 14.5).

Affordability is key to the success of the ADMI Grain Handling System—each dryer cost around US$500 to build and the bags cost approximately US$1.50. Work done on an ADMI-affiliated project in Bangladesh has shown that a famer in Bangladesh can recover the cost of investing in an STR dryer in two years. The premium prices farmers receive for their properly handled grains, combined with the income generated from providing drying services to neighbors, has a substantial impact on farmers' livelihoods (Figure 14.6).

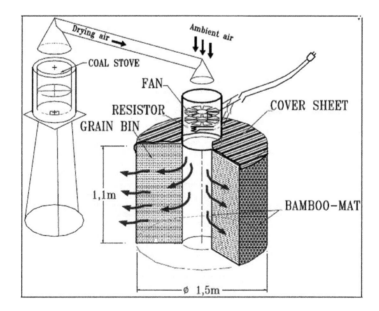

FIGURE 14.5 STR dryer diagram.

FIGURE 14.6 STR dryer training for village farmers.

14.6 PHL PROJECTS IN THE ADMI VILLAGE

14.6.1 Subsidized Bag Distribution

Hermetic bags are a proven, effective technology, but have not been widely adopted in South Asia and only achieve optimal impact when combined with properly dried grains. Despite the many benefits of using hermetic bags, farmers in Bihar lack the resources to spend on products whose value has not been directly demonstrated to them. While hermetic storage bags offer a solution to farmers, they present a financial challenge. The market price for an imported 50 kg hermetic storage bag is approximately INR 95, while the average total per capita expenditure of residents in the ADMI Village is less than INR 50 per day. Purchasing one bag might be feasible, but most farmers would require multiple bags, which would be a real stretch to a household budget—especially for an unfamiliar technology with unknown benefits (Figure 14.7).

In February 2017, ADMI initiated a subsidy program to distribute 21,150 hermetic storage bags in the ADMI Village and 48 additional communities in Bihar. Farmer investment in new technology is essential to ensure proper adoption, so instead of giving away the bags, the ADMI Village provides the bags at a significantly subsidized rate to give smallholder farmers access to the technology. The program gives farmers access to up to five hermetic storage bags at the price of INR 20 each, a discount of more than 75% off market price. The funds generated from the bag sales finance the purchase of additional storage bags to continue the program until farmers are willing to pay market price. Farmers who have purchased the hermetic bags have reported increased annual revenue of more than INR 400 by eliminating pesticide use and reducing crop losses they incur using the jute bags. Hermetic storage bags pay for themselves if they are reused over multiple seasons; an average bag can last three seasons.

14.6.2 Technology Centers

The full value of hermetic bags is only realized if the grain stored in them is properly dried. This means that even if bags are subsidized, farmers in Bihar might not find the improved storage results that warrant the expense unless they also have access to improved drying systems. In

FIGURE 14.7   Storage bag distribution.

2017, four technology centers were established at the ADMI Village in an effort to give farmers access to drying and storage techniques of the ADMI Grain Handling System. The centers are distributed throughout the ADMI Village to provide processing services to a wide range of farmers who travel to the centers on foot with their grain in tow. The centers are managed by trained lead farmers, and are equipped with STR dryers, generators, moisture meters, and hermetic bags. The technology center leaders provide drying services to area farmers, distribute hermetic storage bags, and serve as information sources about proper postharvest management. Both men and women are acting as managers of the technology centers, turning the centers into small agro-enterprises to supplement income from farming activities. The goal of the ADMI Technology Centers is to provide producers with the information and tools necessary to mitigate loss, improve grain quality through adequate drying and storage solutions, and eliminate some of the drudgery associated with traditional drying, which often takes several days. These are new technologies for this area and learning how to use the technology is essential for farmers to get the most return on their investment (Figure 14.8).

14.6.3 Trainings

Training sessions were central in the initial phase of the ADMI Village Project. Several *kisan ghostis* (farmer trainings) were held to demonstrate postharvest technologies and best practices for grain handling. In the second year of the project, the women of the ADMI Village spoke out, requesting trainings for women by women. The FAO estimates that women conduct 60% of postharvest processing activities in Asia (SOFA Team and Doss 2011), so the demand for gender-sensitive trainings came as no surprise. BISA hired a female trainer to provide training and extension services to women at the ADMI Village, as well as villages in the nearby East Champaran District of Bihar. To date, more than 1,500 women have received training. In October 2017, BISA and DRPCAU partnered with the Indian Council of Agricultural Research to organize an event at a local Farm Science Center to mark the International Day of Rural Women. Trainings were held on climate-smart farming systems, including demonstrations of the modified STR dryer, hermetic storage bags, moisture meters, zero energy cooling storage chambers, and solar dryers (Figure 14.9).

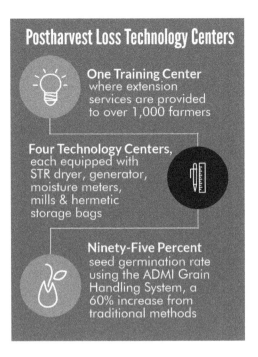

FIGURE 14.8 Technology center chart.

FIGURE 14.9 Women's day activity.

14.6.4 ADMI MOBILE DRYER

The ADMI Village is a site for continued research, innovation and field testing. While the STR dryer has applications for small producers, its size and limited mobility make it difficult for some farmers to access. In an effort to create a faster, more convenient drying solution, BISA introduced a mobile bed dryer in partnership with the ADMI Village. The dryer is built on a set of wheels and tilts horizontally to attach to a tractor for travel. A diesel engine provides the heat for the dryer, while the grain loader is powered by a 2 hp motor and the blower is operated by a 5 hp motor. The mobile dryer can run 1 ton of grain per shift and completes a shift in one to two hours, depending on initial moisture content of the grain. The dryer's combination of higher capacity and ease of transport is promising for smallholder farmers and custom hire service providers alike. The dryer has been piloted in Bihari villages and demonstrated for the Bihar government. BISA is engaging in discussions with the state about an agricultural subsidy program to support large-scale local production of mobile dryers (Figure 14.10).

14.6.5 EXPANDING THE REACH

The focus of the ADMI Village is to provide a space for international conversations and education about PHL solutions. Through this project, ADMI seeks to facilitate interactions between PHL practitioners to create collaboration in research and educational efforts.

In March 2018, the ADMI facilitated a Postharvest Loss Convening at the ADMI Village to bring together research partners from India, Bangladesh, and Illinois to share ideas about technology and discuss the future of PHL prevention in South Asia. Although India and Bangladesh are neighboring countries with similar levels of postharvest challenges, researchers in the two countries rarely interact. At the meeting, Bangladeshi scientists were able to provide their expertise on optimization of the STR dryer design, while BISA demonstrated the ADMI mobile dryer. The researchers made connections for future projects to pool resources and expand the footprint of PHL projects. Connecting practitioners in the field of PHL and promoting economically viable, small-scale solutions are the heart of what ADMI does. Strengthening local partnerships and focusing on green PHL solutions will be key to the future of the ADMI Village project.

FIGURE 14.10 ADMI mobile dryer.

14.7 CONCLUSION

The ADMI Village project is making local impact by providing PHL technology, trainings, extension services, and more to the residents of Dih Sarsauna, Bihar, India. Much of the immediate impact is felt locally, but the project will help lead to a more sustainable and profitable agriculture system in South Asia and around the world. Thanks to daily in-the-village and on-the-farm work, the researchers and villagers continue to learn more about PHL interventions—the technologies that work and are affordable, the awareness that is still needed at all levels about PHL prevention, and the trainings that will help smallholder farmers learn firsthand that PHL is indeed a solvable problem.

REFERENCES

ADMI. (2017). Postharvest technology to improve smallholder income and food security. ADM Institute for the Prevention of Postharvest Loss. Retrieved from https://postharvestinstitute.illinois.edu/featured-projects/postharvest-loss-prevention-in-india/.

Agrilinks. (2018). Food safety hazard: Mycotoxins. Retrieved from https://www.agrilinks.org/sites/default/files/mycotoxins-mm.pdf.

Baylis, K., Pullabhotla, H.K. and Shukla, P. (2016). Reducing postharvest losses in Bihar—Bihar PHL Baseline Survey Report. ADM Institute for the Prevention of Postharvest Loss. March 2016. Retrieved from http://go.illinois.edu/BiharPHLbaseline.

FAO. (2011). *Global Food Losses and Food Waste: Extent Causes and Prevention.* Rome, Italy: FAO. Retrieved from http://www.fao.org/docrep/014/mb060e/mb060e.pdf.

IL&FS (Infrastructure Leasing & Financial Services). (2014). Food processing in Bihar: The road ahead, prepared for the Government of Bihar. Retrieved from http://www.ilfsclusters.com/pdf/Food_Processing_in_Bihar.pdf.

Kumar, D. and Kalita, P. (2017). Reducing postharvest losses during storage of grain crops to strengthen food security in developing countries. *Foods*, 6(1): 8. Retrieved from https://www.ncbi.nlm.nih.gov/pmc/articles/PMC5296677/.

Ranganathan, T. (2014). Farmers' income in India: Evidence from secondary data. Agricultural Economics Research Unit (AERU), Institute of Economic Growth (IEG). Retrieved from http://www.iegindia.org/ardl/Farmer_Incomes_Thiagu_Ranganathan.pdf.

SOFA Team and Doss, C. (2011). The role of women in agriculture, The Food and Agriculture Organization of the United Nations. March 2011. Retrieved from http://www.fao.org/docrep/013/am307e/am307e00.pdf.

YieldWise. (2018). YieldWise food loss—Reducing loss from what we grow and harvest. YieldWise. Retrieved from https://www.rockefellerfoundation.org/our-work/initiatives/yieldwise/.

15 Lessons Learned from a Postharvest Training and Services Center in Arusha, Tanzania

Ngoni Nenguwo, Roseline Marealle, and Radegunda Kessy

CONTENTS

15.1 INTRODUCTION

Most vegetables and fruits should be handled carefully after harvest since they are easily perishable, and poor handling could reduce the quality of the harvested produce, resulting in a lower market price to the producer. Since many fruit and vegetable crops have an inherently short shelf life, they are usually sold immediately after harvest, which may limit the farmer's choices in their marketing activities, resulting in lower incomes. Good postharvest practices usually involve careful handling with appropriate equipment and tools, and use of temperature-controlled storage facilities to increase shelflife of fruits and vegetables after harvest. Smallholder producers of horticultural crops do not always have the proper tools and equipment for postharvest handling and in most cases lack access to facilities such as cold rooms.

A project was developed and initiated to investigate the feasibility of improving access to postharvest equipment and services for smallholder horticultural producers by establishing a postharvest training and services center in Arusha, Tanzania at the World Vegetable Center (WVC) campus. It was proposed that having a wide range of tools and equipment in one location would make it easy for the growers to access and use them. The center would also train farmers in postharvest handling methods to reduce losses in vegetable crops. The present chapter describes in detail the major training activities which took place October 2012–March 2017.

15.2   SUMMARY OF POSTHARVEST TRAINING AND SERVICES CENTER FACILITIES

The Postharvest Training and Services Center (PTSC) was set up with funding from the United States Agency for International Development (USAID) to exhibit the range of tools and equipment that can be used by smallholder growers, traders, and processors while postharvest handling of horticultural produce. Having all the equipment at one place makes it convenient for practical training and for holding demonstrations. The center has a storeroom and a cool room equipped with an air conditioning unit using CoolBot technology (developed by Store It Cold LLC).

Various tools and equipment used for harvesting and grading, such as harvesting bags, grading rings, and grading tables, are exhibited along with packaging and handling equipment, such as boxes and plastic crates, and processing tools and equipment. Cooling demonstrations are provided using the Zero Energy Cool Chamber (ZECC) and CoolBot, and simple processing demonstrations can be made using solar drying equipment.

15.3   SUMMARY OF TRAINING ACTIVITIES CONDUCTED SINCE END OF PROJECT

The following are some of the activities that were carried out by WVC staff after the end of the PTSC project.

15.3.1   Collaboration with Other Organizations

Though the PTSC was hosted at the WVC, its services have been requested by outside institutions such as nongovernmental organizations (NGOs) and organizations involved in developing agriculture and rural livelihoods.

15.3.1.1   Outreach Method

The presence of a dedicated postharvest center can be useful for organizations looking for specific interventions in this field. Over a time, such institutions would become a source of information serving a wide range of customers. The work of the center can become popular through the people working in agricultural development and in advisory services through word of mouth, as well as by attending the functions and activities held at WVC. One such example is the request received from the Agency for Co-operation and Research in Development (ACORD), an international NGO, for the postharvest facilities and expertise that it needed. ACORD approached WVC directly for the help in installing simple cooling facilities at one of their project sites. The organization was working in Geita District, in northwestern Tanzania, and wanted to improve the handling and marketing of pineapples, which are widely grown in the district.

15.3.1.2   Target Group

ACORD had observed that most households in Geita are involved in agriculture, with more than 70% of households dependent exclusively on agriculture for their livelihood. The region has good agroclimatic conditions suitable to produce staple crops, as well as crops for income generation. Many producers in this area produce fruits and vegetables for their consumption, as well as for sale. The organization was involved in improving the production and marketing of fruits and vegetables grown by farmers in this area to increase the income and improve the lives of the inhabitants. ACORD has been providing training and information on agronomic aspects of horticultural crop production. However, the organization had limited knowledge and information on postharvest handling technology, and they had approached the WVC having learnt about the PTSC from previous visits to the WVC campus. An agreement was then drafted for WVC to train ACORD staff in postharvest handling of horticultural crops.

The training was aimed at technical staff of the organization as well as lead farmers and local agricultural extension workers in the region. Representatives from four groups were selected by ACORD for the training, with the understanding that they would further go on to train the producers from other groups. The people selected for this training had a background in agriculture, may have had diploma-level or degree-level training, and included lead farmers with practical experience in fruit and vegetable farming.

Trainees were given theoretical and practical lessons in:

- Introduction to the principles of good postharvest handling and understanding the importance of market requirements
- Packhouse operations (including a practical session on grading and packaging) and options for value addition to the fresh produce
- Cooling and storage requirements and low cost cooling options (including a practical demonstration on how to construct a ZECC)

ACORD had identified the ZECC as a suitable low cost storage method for crops such as pineapple. The ZECC could be used as the produce is collected and bulked up before moving it to markets. The producers in the area also needed training on how to reduce damage to their crops during harvesting and marketing, and how to improve the quality of the crops to be marketed.

The installation of the ZECC was included in the training of ACORD staff, as well as lead farmers, so that they could go on to construct more ZECCs at other sites in the region.

The format for the training was a two-day session with lectures in the morning to explain the theoretical aspects of the topics, and then a practical demonstration in the afternoon to give the trainees hands-on experience and skills. During the training, the participants were encouraged to discuss and share their experiences and knowledge concerning the topics being covered.

General challenges that came up during the training were as follows:

- The trainees indicated that they had observed postharvest losses (PHL) of fruits and vegetables occurring at all stages in the handling chain. The fact that proper solutions had to be implemented by all the players during marketing was mentioned as an important requirement.
- In addition, not all the people involved in the marketing and handling of fruits and vegetables were concerned about PHL; one such example is traders, who were accustomed to damaged produce and consequently factored these losses in the price calculation.
- Producers do not know that effects of poor postharvest handling may not show until the produce has been marketed to others, and the sales of damaged produce could result in losing their customers' good faith.
- An issue of major contention that was raised by the producers was the question of who sets the price on the markets. Most farmers believed that the prices were determined by the buyers and that farmers are price takers and not price setters, and are thus in a weaker position. This reduces incentives to improve postharvest handling and quality of the fruits and vegetables they sell.
- Another problem highlighted by the trainees was that they normally use baskets as containers for their produce, and these baskets are not uniform in terms of volume/quantity, by which producers fail to get deserving income.

The questions raised by the participants showed that it is important to integrate the above points in the training so that all the actors across the value chain can take up the improvements as well. The buyers and sellers, for example, should agree on the quality standards with a higher price paid for better quality fruits and vegetables, a new lesson learned by the trainees.

15.3.1.3 Training Outcomes

Evaluation of the training included an overall scoring exercise for participants to rate the various training tools and knowledge gained. The results were as follows:

- The approach of mixing theoretical lessons and practical exercises was appreciated by the trainees.
- The practical sessions also encouraged discussions amongst the trainees, which was useful as it was one way of helping them share their own experiences.

A mid-term evaluation was conducted 2 years after the training, and the following were the reviews given by the trainees:

- The training was relevant to their needs.
- They wanted refresher courses. This may indicate that some participants did not fully understand some of the topics, which may be due to differing educational levels—a common issue with mixed groups.

Some participants suggested that they needed more time and wanted more knowledge and information on postharvest handling.

The practical sessions were appreciated as a way of helping them to understand and remember what they had been taught.

Since the need was identified to improve postharvest handling methods, three ZECCs were installed during training in the Geita region of Mwanza, Tanzania and ACORD staff went on to construct another ZECC soon after. These low cost cool storage structures are being used for storage of pineapples since ACORD had identified this value chain as being important for the producers in this area. A follow-up visit made a year after the training has found that an additional 15 ZECCs had been constructed and were being used by producers' groups working with ACORD in the region.

Organizations such as ACORD have used postharvest training as a necessary activity to complete their value chain interventions and have sought the input of the PTSC in providing postharvest technical information that their staff may not have.

15.3.2 Farmers and Farmer Groups

Many of the users of the PTSC were farmers/farmer groups, and it was worthwhile to encourage them to approach the center directly for training. The center could suggest appropriate techniques and technologies to mitigate their postharvest handling problems and exploit new opportunities previously unknown to the farmers/farmer groups.

15.3.2.1 Outreach Method

Advertising activities and services of the center through open days and word-of-mouth efforts can result in reaching a wide range of actors, including farmer groups and the representatives of farmer organizations, and there exists the possibility to get an idea about the specific information or training expected by them. During the 2-year period of the PTSC project, many people who came for training at the center were selected from farmer groups in and around the Arusha region of Tanzania, due to their proximity to the center. However, some of these trainees provided positive feedback to their associations and members within their groups, and thus the information about the postharvest training provided by the PTSC reached other farmer groups by word-of-mouth. As a result, information about the PTSC has been spreading amongst different sectors involved in fruit and vegetable production. These included Mtandao wa Vikundi vya Wakulima Tanzania (MVIWATA) in Swahili, or Association of Farmers Groups in Tanzania in English. They subsequently requested training for members in other regions.

15.3.2.2   Target Group

MVIWATA officials from Morogoro and from Lushoto regions of Tanzania requested training for their members in postharvest handling of vegetable crops, including an emphasis on methods suitable for smallholder producers, such as evaporative cooled storage structures (for example, the ZECC). Many of the farmers in Morogoro and Lushoto are involved in the production of vegetables and market them to the major towns and urban centers. Hence they were interested in techniques to improve quality of their produce.

The training conducted included practical sessions on the handling of vegetables, including grading exercises. This aspect of the training raised a lot of discussion from the farmers about the benefits of grading their crops prior to packing and selling. Grading is useful where the grower was able to identify different markets for different quality products. Other producers, however, believed it was up to the buyer to do the grading, packing, and transportation.

15.3.2.3   Training Outcomes

Subsequent follow-up with MVIWATA in Morogoro revealed that they had constructed a ZECC as a demonstration, and three others were constructed by their members—a total of four ZECCs that they had constructed after the training.

An example of training outcomes is farmer to farmer promotion. Some of the participants of the postharvest training are willing and able to share their knowledge and skills with their neighbours. A good example is that of Mr. Sospeter from Morogoro. In January 2016, Mr. Sospeter and other members of his group constructed a ZECC and then held a field day, inviting farmers to attend and witness how it functions. A total of 20 farmers attended the field day. Following this, two more ZECCs were installed at Chanzema and at Lolo-Langali areas by the farmers who attended the field day, and were linked to MVIWATA (Mtandao wa Vikundi vya Wakulima Tanzania), which is a network of farmers groups in Tanzania. In addition, the District Agricultural Officer decided to construct one at the Nanenane Agricultural Show in Morogoro to demonstrate this technology to people who passed through their booth. MVIWATA farmers group have mentioned that they are planning to construct one in the local market as a way of further promoting the technology.

The link with groups such as MVIWATA was useful in reaching a larger number of farmers; additionally, they could bring real-world problems to the center. A major weakness of the work with some of these farmer groups is that the PTSC may not have funds for follow-up support, and will have to depend on the farmers' association for their own initiative in further promotion of the technologies.

15.3.3   Lessons for Hosting Institutions

15.3.3.1   Outreach Method

The PTSC was set up with emphasis on appropriate low cost technologies suitable for smallholder farmers. This was useful for WVC, which, as the hosting institution, could include postharvest activities in its program.

15.3.3.2   Target Group 1

One example was the Vegetables for Income and Nutrition in Eastern and Southern Africa (VINESA) project to establish best practice hubs, which was funded by Australian Centre for International Agricultural Research (ACIAR). The project was aimed at training youth in improved vegetable production technologies. This project was implemented using a value chain approach to the trainees learning how to identify the market requirements in terms of quality for vegetables of greatest interest to them. The overall training program was six months long, covering all aspects of vegetable production, postharvest handling, and marketing. Selected postharvest technologies were included to help the trainees improve the quality of their produce. The identified techniques included the use of grading tables to make grading easier and faster, so helping the young producers to meet the quality specifications as per the market needs. The trainees were taught the use of good quality grading tables that can be disinfected (tables made from stainless steel, rather than wood). Another selected technology was the ZECC, which was included as a way of increasing the shelf life and reducing deterioration of vegetables supplied. The trainees were also introduced to high-quality plastic crates, rather than commonly used wood crates. These wood crates have rough surfaces that cause damage to the packed produce.

15.3.3.3   Training Outcomes

The trainees appreciated the various technologies and the requirements for improving the quality of the crops they sell. One group of trained youth was able to identify a site for the introduction of a satellite hub, where they established a collection center for produce that had a grading table and ZECC, as per their request. However, funds for these facilities had to be provided by the project. Having a center such as the PTSC could, therefore, be used as the hosting center, strengthening the range and scope of services and technologies the institution promotes. Groups that have regular access to the PTSC are farmers and producer groups around Arusha, trainees from training institutions, technical support staff, and other beneficiaries from projects managed by WVC staff.

The PTSC was a useful component of the training activities of the ACIAR-funded VINESA project, which set up best practice hubs in vegetable production zones to train the local youth in production and marketing of high value crops. Identification of value chain opportunities and the requirements for improving quality to meet these opportunities was undertaken during the implementation of the VINESA project.

15.3.3.4   Target Group 2

Another group of trainees brought to the PTSC by WVC staff were farmer and trader groups growing vegetables for marketing in major urban centers in Tanzania and Kenya. These farmers were selected for training under the WVC Postharvest Program Component, a postharvest project funded by USAID. The project promoted improving postharvest handling and reducing losses of vegetable crops by introducing new postharvest technologies. Farmer groups identified were involved in tomato production in Arusha, Lushoto, and Morogoro regions of Tanzania. Training sessions and demonstrations were provided using two-day courses for improved postharvest handling, improved packaging, and low cost cooling methods (the ZECC and CoolBot at the PTSC).

The training provided comprised lecture sessions in the morning and practical exercises in the afternoon. The participants were able to view a full range of postharvest technologies being promoted, starting from harvesting aids, grading tools, packaging materials, and processing materials, as well as low cost cooling equipment such as shade structures, the ZECC, and the CoolBot controlled cool room (Figure 15.1).

15.3.3.5   Training Outcomes

Some of the training practices included giving the trainees problem-solving scenarios that they may encounter, asking them to use their new postharvest information to develop solutions. Problem-solving was used as a way of encouraging discussions among the participants and as a learning tool. Typical problems given to the trainees included:

a. If you have identified providing vegetables to a local hospital as a market opportunity, what grading criteria could you use for tomatoes and for leafy vegetables?
b. If you have identified a market at a local tomato processing factory, how would you grade and store the tomatoes?

FIGURE 15.1   Trainees discussing ZECC.

c. If your transporter has informed you that the vehicle has broken down on the way to pick up your produce, what method of storage can you use while waiting for the vehicle to be repaired?

d. If you have found a demand for your tomatoes at a high-end supermarket, but the buyers are complaining of damaged fruit, how would improve the quality of the crop?

Observations made by the trainees during and after the training included:

a. Grading criteria for hospitals
   • Produce must be fully ripe since they do not want to store it
   • A problem is delayed payment, so sometimes farmers are reluctant to supply
   • Grading criteria for size is not important, since produce will be used for cooking
b. Grading criteria for tomato processing factory
   • Produce must be fully ripe
   • Size of fruit not important
   • Disease and blemishes not good (fruit will be rejected)
   • Some trainees suggested that larger fruit to be sold to the fresh market and smaller fruit to factory
c. Temporary storage method used while waiting for the vehicle
   • The produce can be placed in a cool environment or shade
   • Some suggestions made to sprinkle water over the produce or to cover the boxes with leaves
d. Problem of damage to tomatoes
   The trainees listed the sources of damage as due to:
   • Poor field management (lack of enough water for irrigation, pests and diseases)
   • Late harvesting of the crop resulting in overripening
   • Poor handling during harvesting stage
   • Compaction of tomatoes in the crate

The trainees proposed the following measures to reduce losses in tomato crops
- Training casual workers to be careful while handling the produce
- Avoid allowing sun to shine directly on harvested tomatoes
- Improve the packaging materials
- Improve field management practices

The overall conclusion from participants from the problem-solving exercises:

1. Grading of vegetables can lead to better returns
2. It is always important to understand what customers want
3. It is important to understand demand and supply forces in the market, as well as purchasing power of customers (Figure 15.2)

An evaluation of the USAID-funded postharvest project was carried out 2 years later to determine the effectiveness of the training; some of the findings and criticisms included:

- **Short duration of training and limited follow-up support**. Some of the trainees who were interviewed felt the training duration was too short and lacked a structured follow-up. They would have liked support after the training as a refresher course and to help them further spread the message of improved postharvest handling.
- **Alternative solutions for lower quality produce**. Outlets for substandard fruits and vegetables need to be identified. Good examples are tomato seed being produced by the seed companies, and the processing market, especially when there is no market for the fresh produce.
- **Focusing on the postharvest part of the value chain may not resolve all issues faced by producers**. Although the technologies were intended to resolve postharvest problems faced by the farmers, many of them had problems with preharvest issues, such as pests and diseases. The result was that the upstream areas of the value chain are not addressed in a systematic manner.
- **Value chain collaboration needed**. The staff of the PTSC may not have the resources to provide a full complement of support and advice along the value chain; consequently,

FIGURE 15.2   Trainees discussing packaging material.

it would be useful to link trainees with other organizations and actors to fill in the gaps. Many farmers, for example, requested help with accessing markets and linkages with other organizations that could provide some of these services.

Many of the trainees had expressed general satisfaction with the training process and proceeded to use some of the lessons from the postharvest course. It is interesting to note that the technologies usually implemented were the use of shade and the improved grading practices. They did not set up their own CoolBot controlled cold rooms, which may be due to the high establishment and maintenance cost of the technology. It would be useful, therefore, to link the training provided in the PTSC with sources of support or funding to help the trainees take up the more expensive technologies.

A weakness observed was that there was no structured follow-up activity to encourage trainees to pass on the information they have learnt to others. However, it was noted that some trainees could communicate with approximately 7–12 of their neighbors and colleagues.

15.3.4 SUMMARY OF OTHER OUTREACH ACTIVITIES

Promotion of the activities and services provided by the PTSC was made through various means, such as when the WVC hosted its open days or attended local agricultural fairs. The advantage of this outreach method was that a wider audience and range of actors involved in the horticultural value chain could be reached. Especially, it was a useful way to reach out the private sector actors for potential collaboration and linkages.

The radio, television, and public media were also used as a way of promoting the work of the PTSC. Several interviews for radio and television in Tanzania were attended by staff working at the PTSC, who could effectively publicize the benefits of improved postharvest handling methods, such as the use of good quality plastic crates instead of the wood crates, which are normally used by producers and traders.

Some of the postharvest tools and equipment used at the PTSC have been taken to various fairs to demonstrate alternative ways of informing and reaching out to potential users, as well as to promote these tools. Brochures on the technologies have been made available for visitors to take away, and videos are shown to demonstrate the technologies in action. Agricultural shows at the local level include shows such as the Meru Agricultural Fair, an annual agricultural show held in the Meru District of northern Tanzania, which is attended by approximately 500 people. Larger shows include the Nane Nane show, an annual agricultural show held in August at national and provincial levels, which is visited by the public. Up to 1,500 visitors came to the WVC booth at the show in Arusha, Tanzania.

These events allow exposure of PTSC activities to a wider audience, although the outcome/adoption is not easy to measure. However, some of the anecdotal evidence shows that some members of the public do pick up the techniques and technologies. In one instance, samples of harvesting bags on display interested the manager of a coffee plantation, who came back to get more information on how he could make his own bags for use on his estate.

15.4 CONCLUSIONS

The results of the training activities carried out under the PTSC show that improved handling techniques, which required low or no financial outlay, were readily adopted. While there was some appreciation from smallholder producers of horticultural crops for the various postharvest technologies being promoted by the center, follow-up surveys indicated that the beneficiaries would have preferred sessions that were longer in duration and more numerous. The most likely chance of adoption of the postharvest technologies occurred when collaboration with other organizations took place that provided a holistic value chain approach, including market access and financial support for more costly technologies, such as cooling equipment.

BIBLIOGRAPHY

Acedo Jr. A., Weinberger, K., Holmer, R. J. and d'Arros Hughes, J. (2012). Postharvest research, education and extension: AVRDC-The World Vegetable Center's experience. *Acta Horticulture*, 943, 47–61.

Bingen, R. J. and Simpson, B. M. (2015). Farmer organizations and modernizing extension and advisory services: A framework and reflection on cases from Sub-Saharan Africa, MEAS Discussion Paper.

Chowa, C., Garforth, C. and Cardey, S. (2013). Farmer experience of pluralistic agricultural extension, Malawi. *The Journal of Agricultural Education and Extension*, 19(2), 147–166.

Kitinoja, L. (2013). Innovative small-scale postharvest technologies for reducing losses in horticultural crops. *Ethiopian Journal of Applied Science and Technology*, 1, 9–15.

Kitinoja, L. and Barrett, D. M. (2015). Extension of small-scale postharvest horticulture technologies—A model training and services center. *Agriculture*, 5, 441–455.

McNamara, P. E. and Tata, J. S. (2015). Principles of designing and implementing agricultural extension programs for reducing post-harvest loss. *Agriculture*, 5, 1035–1046.

Swanson, B. E. (2008). *Global Review of Good Agricultural Extension and Advisory Service Practices*. Food and Agricultural Organization: Rome, Italy.

16 Farmer Advisory Services in Tajikistan (FAST)

Agriculture Extension Program to Prevent Postharvest Losses and Improve Nutrition

Lola Gaparova and Andrea B. Bohn

CONTENTS

16.1 FAST AND THE FEED THE FUTURE/TAJIKISTAN STRATEGY

16.1.1 INTRODUCTION TO THE PROGRAM

This book chapter is based on the final evaluation report of the US Agency for International Development (USAID) funded and University of Illinois Urbana-Champaign (UIUC) implemented project entitled Farmer Advisory Services in Tajikistan (FAST). FAST was mainly responsible for building the capacity of local institutions and community-based organizations, and completion of effective agrarian reform in selected districts of Khatlon Province, Tajikistan. FAST was responsible for developing an Extension and Advisory Services (EAS) model targeting small commercial and household farmers (who are predominantly female), which would be sustainable beyond the project life, and therefore contribute towards improved rural livelihoods, household food security, better nutrition, and prevention of postharvest losses (PHL). The FAST project provided special attention to gender roles in the design of the EAS and, as such, deliberately took a participatory approach to extension services to include the opinions and experiences of the women farmers.

This chapter provides a summary of the final months of the project's extended field operations (October–December 2015), specifically on the work that FAST completed with the Household Farm Learning Groups (HFLGs) in cooperation with USAID's Farmer-to-Farmer program, production of mass media training videos, and other noteworthy technical activities. The chapter mentions success stories and includes the data collected on the second season crops (corn and mung beans, called *mosh* in Tajik). Finally, the chapter includes data/information from the FAST Program Final Performance Report (Modernizing Extension 2015).[1]

UIUC implemented FAST under a Leader with Associate Cooperative Agreement from USAID. The Modernizing Extension and Advisory Services (MEAS) consortium, based at UIUC and implemented under a separate cooperative agreement, was the "leader" in this award.

16.1.2 LOCATION OF PROJECT

The project was successful in establishing an effective EAS in 11 of the 12 USAID designated Feed the Future (FTF) districts in Khatlon province in southern Tajikistan (Note: the 12th district was not agriculturally active and therefore excluded) (Figure 16.1). Within a period of 2.5 years, the project reached 3,898 mainly female household farmers through establishment of HFLGs. In addition, 828 small commercial farmers, Water Users Associations (WUA), and local government officials were also trained.

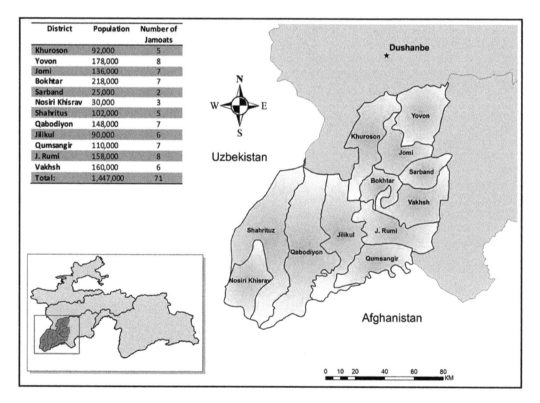

District	Population	Number of Jamoats
Khuroson	92,000	5
Yovon	178,000	8
Jomi	136,000	7
Bokhtar	218,000	7
Sarband	25,000	2
Nosiri Khisrav	30,000	3
Shahritus	102,000	5
Qabodiyon	148,000	7
Jilikul	90,000	6
Qumsangir	110,000	7
J. Rumi	158,000	8
Vakhsh	160,000	6
Total:	1,447,000	71

FIGURE 16.1 Map of 12 FTF districts in Tajikistan and overview of population and number of *Jamoats* (sub-districts) per district (2014).

---

[1] The final performance evaluation was conducted by an external evaluator retained by the program's implementer, UIUC.

16.1.3 Aims and Objectives

The objective of the activity was to contribute to the achievement of USAID's Feed the Future/ Tajikistan (FTF/T) strategic goals of sustainably reducing poverty and strengthening food security in the Republic of Tajikistan.

For achieving project aims and objectives, the following extension models were identified: (1) use of participatory methods to determine the farmers' problems and needs, (2) organize female farmers into HFLGs, (3) develop a workshop/training program based on the local crop calendar, (4) implement practical and field-based training with a focus on cost saving and yield increasing technologies, (5) assign technical specialists and a group facilitator to each learning group and promote their collaboration, and (6) pursue technical representation of the project at the subdistrict (*jamoat*) level through the Jamoat Extension Coordinator (Figure 16.2).

16.1.4 Strategy

16.1.4.1 The Three Pillars of the FTF/T Strategy

Pillar I: Assistance to household and small commercial farms to increase income and food production for home consumption, and to improve nutritional and health outcomes
Pillar II: Building the capacity of local institutions and community-based organizations
Pillar III: Completion of effective agrarian reform in selected districts of Khatlon Province

FAST supported and ensured the implementation of USAID's FTF strategy in the Republic of Tajikistan. The approved results framework in its Project Monitoring Plan indicates that the goals of these three pillars are: *Pillar I*: to design, develop, prototype, and document an EAS that is capable of reaching smallholder farmers, particularly household farmers; *Pillar II*: to transfer that system to a follow-on project to be rolled out throughout the Zone of Influence (ZoI); and *Pillar III*: to provide policy and capacity building support to the Government of the Republic of Tajikistan (GoTJ) to implement integrated agrarian reforms.

The FTF/T strategy began from the observation that most citizens of Tajikistan are dependent on production from their household farms and related small commercial farms for their subsistence. They consume much of what they produce and sell the balance to obtain cash and additional

FIGURE 16.2   Tajik farmers at a women's group meeting.

foodstuffs. Therefore, the primary task of improving food security in Tajikistan is to help those smallholders produce more food and of higher nutritional content for consumption and sale. As they increase and improve their output, and as the ongoing agrarian reform in Tajikistan make mores resources available to them, many smallholders were expected to develop a more commercial orientation, increasingly producing for the market.

The FTF/T strategy was focused on 12 administrative districts in the Qurghonteppa zone of Khatlon province. This area was the FTF/T ZoI. The 12 districts were subdivided into 76 territories governed by councils (*jamoats*), of which 71 were entirely or partially rural.[2] The territory of each rural council included one or more villages. Villages are recognized territorial entities. The neighborhood (*mahalla*) is part or all of a village, is usually centered near a mosque or other community building, and is the lowest level administrative unit.[3] There are 727 remaining state-owned farms, 22,743 *dehqon (or dekhon)* farms (dehqon means private owned/farmer farm), and about 180,000 rural household farms in the FTF/T ZoI. About 40% of the *dehqon* farms hold less than two hectares of arable land. About 60% of the farms hold less than five hectares of arable land. Therefore, about 10,000 of the *dehqon* farms are "smallholder," as defined in as FTF/T target indicators. In total, at present, there are almost 200,000 smallholder households and commercial farms in the ZoI.

Almost all arable land in the ZoI is irrigated. The GoTJ has established WUAs to manage primary, secondary, and tertiary irrigation and drainage facilities throughout the ZoI. Many of those WUAs are not operating productively due to a number of institutional development and financial sustainability considerations. However, USAID continues its long-term effort to create and support effective WUAs in the ZoI (Statistical Agency under the President 2013).

Until 1991, essentially all local social services, including education and medical care, as well as functions analogous to agricultural extension in a market economy, were provided to rural residents through the collective and state farms. The ongoing land reform led to the dissolution of the Soviet-era large farms. Their land and remaining productive assets have been distributed to smaller farmers, but without much agricultural production support for rural residents. This situation, together with the country's precarious food security, is the backdrop for the FTF/T strategy.

Of the three pillars of the FTF strategy, Pillar I directly aims to improve nutrition, production, and incomes of small farm households. FAST was, essentially, responsible for the implementation of Pillars II and III. If the activities under those pillars were not successful, then Pillar I efforts cannot succeed. For Pillar II, FAST designed, piloted, and supported the development of a pluralistic, participatory agricultural EAS to assist household and small commercial farm development. Once developed, this EAS was to be handed over to a new project that could roll it out throughout the ZoI. For Pillar III, FAST provided continuing policy analysis and implementation support to the GoTJ to complete critical land, water, and policy reforms to allow for innovation and development of a productive agriculture sector.

The United States Government, through USAID, invested $5.5 million in FAST.

16.2 FAST ACTIVITIES

16.2.1 HFLGs in Cooperation with USAID's FTF Program

During October 2015, FAST continued its final month of extension activities and assisted existing HFLGs. The final standard FAST extension task for each HFLG during each growing season is an analysis of crop production results, referred to as an "end of season evaluation workshop" (ESEW). During October, ESEWs were conducted with 59 groups of women to evaluate the performance of the second season production of corn and *mosh*.

---

[2] Five are settlement *jamoats* without any rural hinterland—they have no territory or authority outside the town boundary.

[3] The entire land surface of Tajikistan, except for cities and bodies of water, is divided up into *jamoats*. However, since *mahallas* are areas where people live, there are no *mahallas* outside inhabited areas.

TABLE 16.1
Corn Statistics

Corn: average for 59 groups (140 learning plots)	2015
Yield, kg/*sotka*	76
Expenses, TJS/*sotka*	32
Gross income, TJS/*sotka*	156
Net income, TJS/*sotka*	124
% harvest sold	24%
% harvest consumed in household	76%

Note: 1 *sotka* is 100 m².

Average consumption rates and income earned with corn (*Zea mays*) are presented in Table 16.1. The second season corn was grown on 140 learning plots (approximately 100 m² each) in 59 HFLGs. The gross harvest on each household farm where FAST had a second season corn learning plot (1 *sotka* = 100 m²) averaged 76 kg. Average net income per *sotka* is TJS 124. On average, the lead farmer's household sold only about 24% of the harvest, and consumed or gave away to other members of the group, relatives, and neighbors to consume more than 76% of its total corn production. Corn is mainly used for livestock feed, but is consumed in households in the form of green corncob (silage—the crushed green mass of herbaceous plants suitable for animal and bird feed). Expenses include the price of seed and any other inputs provided by FAST. All monetary values are nominal ones. Expenses do not include the value of the labor of the household farmers and their families who hosted the learning plots, or the other members of the learning group who may have worked on the plot. Income is calculated at the market price for the product as stated by the group participants and assumes that the entire amount was sold.

During the second crop season in 28 learning plots in 48 HFLGs *mosh* was cultivated as a high nutrition (protein) crop. The gross harvest on each household farm where FAST had 48 learning plots (1 *sotka* each) averaged 27 kg. On average, the lead farmer's household sold about 28% of its harvest and consumed or gave away to other members of the group, relatives, and neighbors to consume 72% of its total production. The net income per *sotka* is 189 TJS. The crop prices were comparatively low, due to the abundant supply that year; however, results were reasonable even in a low price year. The other important topic covered during HFLG at the end of season discussion was vegetable and fruit storage technologies under household conditions to prevent postharvest losses (PHL) (Table 16.2; Figure 16.3).

The results of field activities showed that the majority of farmers lost a lot of produce during or after harvest. The FAST program targeted broader stages of the supply chain, such as: harvesting, drying and

TABLE 16.2
Mung Bean Statistics

Mung bean: average for 28 groups (48 learning plots)	2015
Yield, kg/*sotka*	27
Expenses, TJS/*sotka*	27
Gross income, TJS/*sotka*	215
Net income, TJS/*sotka*	189
% harvest sold	28%
% harvest consumed in household	72%

Note: 1 *sotka* is 100 m².

FIGURE 16.3   Tajik farmer with mung bean harvest © FAST. (Courtesy of P. Ludgate, 2015.)

storage, transportation, processing, and retail. To provide the best postharvest technologies to farmers, the program closely cooperated with Agricultural Cooperative Development International/Volunteers in Cooperative Assistance (ACDI/VOCA), which implements the 5-year USAID-funded Farmer-to-Farmer (F2F) program in the Europe, Caucasus, and Central Asia regions (F2F-ECCA). The project's core countries include Armenia, Georgia, Kyrgyzstan, and Tajikistan (Satybaldin 2001).

The F2F program assisted Tajik farmers in making informed agricultural investments and improving their fruit and vegetable production and productivity. The F2F volunteers are from the United States and provided valuable short-term (2–3 weeks) training to assist rural enterprises, producer organizations, and individual farmers. FAST, in cooperation with the F2F program, provided women and small commercial farmers with access to modern agricultural production technology, extension services, and, most importantly, met the primary task of improving food security in Tajikistan by helping the farmers produce more and better, more nutritious products for their household consumption and for sale. A total of 17 training sessions on preharvest and PHL issues were held during this month in which 337 farmers participated.

During the 2 years of FAST and F2F collaboration, the two programs' efforts significantly aided the household and small commercial farmers in Khatlon province by providing 164 training programs to 4,204 farmers through their joint training programs for production, business management, food canning and preservation, and storage in winter.

16.2.2   FAST Success Stories

Over the 26 months of the project, FAST submitted 16 success stories to USAID and developed a training manual in farmer advisory services (Van Atta et al. 2015). Crop-focused booklets and tip sheets, including improved postharvest handling practices, were developed for 10 crops. These success stories include a final five success stories during the months of October and November 2015 entitled:

1. "USAID training program provides third crop production"
2. "USAID trains rural women to produce nutritious crops"
3. "Collaboration increases agricultural production in Khatlon"
4. "Drip irrigation system reduces household expenses"
5. "New skills increase the yield of sweet potatoes"

16.2.3 DEVELOPING EDUCATIONAL TRAINING VIDEOS

The project's training video editor finalized seven training videos and copied them onto DVDs for USAID's review and distribution to FAST household farmers and USAID's partnering programs.

The first set of videos is for orchard management and were created with the help of one of the F2F volunteers. The videos include explanations/commentary by a local horticultural specialist and are only available in the Tajik language.

The second CD holds four videos covering various aspects and methods for food preservation/canning and were produced in both Tajik and Uzbek languages by a local food preservation specialist. All training videos include topics on preharvest and PHL prevention technologies. They were distributed to 400 household farmers for sharing among their neighbors and friends. The F2F program continued to provide household farmers with information on food preservation and canning using these videos during the following months, in preparation for the spring harvests.

16.2.4 PRESENTATION AT THE FIRST INTERNATIONAL CONGRESS ON POSTHARVEST LOSS PREVENTION IN ROME, ITALY

The First International Congress on Postharvest Loss Prevention held in Rome from October 4–7, 2015 was a very successful, first-of-its-kind, high-level forum on PHL prevention. FAST was represented at the congress by the author, Lola Gaparova, Senior Extension Coordinator. Participants from 64 countries were engaged with the ADM Institute for the Prevention of Postharvest Loss for a program called Developing Measurement Approaches and Intervention Strategies for Smallholders. The program was enriched with 8 keynote speakers, 42 oral presentations, and 78 poster presentations.

The congress assessed the challenges associated with PHL within the framework of metrics and measurements. The focus was to enable the development of better tools and interventions to prevent PHL for smallholder farmers in developing countries. The event provided excellent networking opportunities for professionals from academic, government, non-governmental organizations, foundations, and private entities. This high-level coalition of diverse professionals created a roadmap for PHL prevention by formulating needs and plans for future actions towards a global consensus on measurement and mitigation approaches (Figure 16.4).

FIGURE 16.4 First International Congress on PHL: William Lanier (left) and Lola Gaparova (right) during the poster session at the PHL Congress.

16.3   FAST EVALUATION AND REPORT

16.3.1   COLLABORATION WITH THE F2F PROGRAM

FAST continued the fruitful cooperation with USAID's F2F program with one-off training events on orchard management, business planning, vegetable production, food canning and preservation, drying, and storage. Farmers often have difficulty obtaining seeds, fertilizer, and pesticides. Farmers are often unfamiliar with modern business management practices. PHL are significant due to antiquated or nonexistent transportation equipment and a lack of climate-controlled storage facilities. The majority of domestically harvested food is consumed through the fresh market, leaving food processing plants to operate at a fraction of their capacity. Quality control, certification, marketing, packaging, and branding of Tajik products is weak.

FAST extension staff was in charge of planning and organizing the workshops, informing and collecting the participants, and facilitating the process throughout the year. The total number of training sessions and participants as the result of FAST and F2F cooperation during the fiscal year (FY) is provided in Tables 16.3 and 16.4.

The numbers of farmers trained are evidence of the reliability and flexibility of the FAST extension and advisory model, and the excellent facilitation and management skills of the F2F program manager and the extension trainers.

Both programs' strong linkages with the farmers and communities in the project areas allowed comparative ease in mobilizing the farmers, and ensured that messages (good agriculture practices, information about new postharvest technologies, marketing strategies, and crop/fruit varieties, etc.) from the F2F volunteer partners would be effectively delivered to the target beneficiaries (household and small commercial farmers).

The USAID agricultural and water development program that followed FAST were advised to capitalize on the training resources provided through F2F. The 3,867 farmers trained during the last year of the program were evidence of the reliability and flexibility of the FAST extension and advisory model and the excellent facilitation, management, and training skills of the F2F

TABLE 16.3
Training Sessions Conducted

Periods	Number of Training Sessions	Number of Participant Farmers
FY 2014 (September 2014)	20	279
FY 2015	127	3,588
Totals	**147**	**3,867**

Note: FY means United States Federal Fiscal Year, from October 1 to September 30.

TABLE 16.4
Increased Crop Yield and Production

Crop	Number of Learning Plots	Yield, in tons/ha 2015	Yield, in tons/ha 2014	Yield Increase in 2015 Compared 2014, %
Potato	158	33.9	19.6	72%
Onion	51	38.4	15.9	141%
Carrot	6	39.0	34.9	12%
Cabbage	6	25.1	24.4	3%
Wheat	3	5.0	3.9	29%

program staff. Early planning and scheduling for potential F2F volunteer assignments, including seasonal timings, locations, and subject matter specialties, would have resulted in optimal utilization of this USAID resource.

16.3.2 Working with Small and Big Commercial Farms

Almost all private commercial farms are, at least formally, administered by men. Most of them have, at least until very recently, concentrated on commodity field crops—cotton and wheat, and more recently, in many cases, onions, carrots, and melons. As the GoTJ campaign to replace cotton with horticultural crops and orchards has unfolded during the last 5 years, this crop mix has changed, but not consistently or everywhere in the ZoI. In many cases commercial farms lease out their land to households for horticultural crop production after the farm's main season is over.

During the last quarter (July–August–September 2015), FAST continued to work with groups of small commercial farmers in selected *jamoats* where the EAS was already working with household farmers. In total, during the 2015 fiscal year, 20 training sessions on agriculture production subjects were conducted in which 446 small commercial farmers participated. Additionally, 525 representatives from commercial farms and WUA participated in the trainings with local officials on issues of agriculture taxation, crop diversification, and input supply issues, as well as domestic and foreign market development. The target for FY 2015 was to pilot EAS for small commercial farms and train 200 farmers in 10 farmer groups. Totally, 971 farmers (producers) were trained, which is almost four times more than initial targets.

In a synthesis of the training sessions Question and Answer discussions, small commercial farmers wanted training in:

- Selection and purchase of good quality vegetable seeds
- Effective practices for storage (preservation) of fruits, vegetables, and beans
- Proper crop rotation systems
- Use of fertilizers on horticultural crops
- Pest and plant disease identification and use of commercial and homemade crop protection preparations
- Integrated pest management

These were generally the same topics pursued with HFLGs. There is clearly demand for a "farmer field school approach" in working with small commercial farmers. In many ways, they are an easier audience if the trainers are properly prepared and experienced, since they provide their own fields without requiring any project contribution for inputs, are more articulate in expressing their desires for training and production/marketing information, and have cash to purchase required inputs such as fertilizers and disease or pest control remedies.

16.3.3 Increasing Nutrition: Results in FTF Extension Programs

Improving the nutritional status of mothers and young children was one of the principal objectives of FTF, a reflection of the growing recognition of the importance of nutrition in childhood development and of the interrelationship of agricultural production, incomes, and childhood nutrition. As a result, there was increased attention on the use of EAS to deliver nutrition messaging and improve nutritional outcomes. The next step that the FAST follow-on program should consider is methodologies to focus attention on integrating nutrition into FAST (farmer field school-like model) by integrated agriculture-nutrition extension services in the ZoI, thereby increasing production and consumption of diversified macronutrient-rich foods. This can be accomplished by adapting the FAST (EAS) tools, such as farmer field schools, to strengthen the linkages of agricultural production to nutrition education and training, and demonstrating positive outcomes, especially among poor farmers.

The follow-on program to FAST was advised to consider working collaboratively with the USAID funded program Integrating Gender and Nutrition in Agricultural Extension Services (INGENAES), which was designed to assist, USAID's FTF missions to:

1. Build more robust, gender-responsive, and nutrition-sensitive institutions, projects, and programs capable of assessing and responding to the needs of both men and women farmers through EAS
2. Identify, test the efficacy of, and scale proven mechanisms for delivering improved EAS to women farmers
3. Disseminate gender-appropriate and nutrition-enhancing technologies and access to inputs to improve women's agricultural productivity and enhance household nutrition
4. Apply effective, nutrition-sensitive, extension approaches and tools for engaging both men and women

16.4 CONCLUSION

An external evaluation of the USAID-funded, and UIUC-implemented project, FAST, was carried out through field work by the consultant in July and August 2015. The project life covered the period July 18, 2013 until September 30, 2015. The project was successful in establishing an effective EAS in 11 USAID designated FTF districts in Khatlon province in southern Tajikistan (Feed the Future 2015). The EAS model developed by FAST is suitable for scaling up, provided that some cost-saving modifications are included. For further roll out of the EAS model, political and financial support from the GoTJ is required. The EAS has been developed for household and small commercial farming, aiming to produce enough nutritious food for the family while selling some surplus.

FAST has been banking on attracting future government support and budget allocations, but did not include alternative options for up-scaling, such as the involvement of private extension providers, farmer/community organizations, or by giving a bigger role to the group leaders of the HFLGs. The project faced the challenge of recruitment of suitable female field staff to work with and the availability of short-term technical assistance. The project provided considerable resources for intensive training and coaching of key project staff, to bring them up to par with the required participatory attitudes and modern technical production skills. Through learning by doing, the project had to find out what positions, and the minimum numbers of staff, that were needed to implement the EAS. Although activity implementation at the field level progressed well, the project set-up and communications between project partners was not ideal.

In conclusion, FAST could develop an agricultural extension model with the potential to reach a larger number of small commercial and rural households in Tajikistan, contribute towards rural livelihood improvement, household food security, and empowering rural women.

This chapter is based on learnings from the United States Agency for International Development (USAID) and US Government Feed the Future FAST project. USAID is the leading American government agency building social and economic prosperity together with the government and people of Tajikistan. The work was made possible by the generous support of the American people through USAID. The contents are the responsibility of the authors and do not necessarily reflect the views of USAID or the United States government.

REFERENCES

Feed the Future. (2015). Sustainable composting provides valuable soil improvements and household income for women farmers in Tajikistan. Retrieved from http://feedthefuture.gov/article/sustainable-composting-provides-valuable-soil-improvements-and-household-income-women.

Modernizing Extension and Advisory Services. (2015). Tajikistan—FAST Program. Retrieved from www.meas-extension.org/home/associate-awards/farmer-advisory-services-tajikistan-program-fast.

Satybaldin, A. (2001). Post-harvest development status and opportunities in the countries of Central Asia and Caucasus. Retrieved from www.agroweb.unesco.kz/fromhom/regrep.html.

Statistical Agency under the President of the Republic of Tajikistan (SA), Ministry of Health [Tajikistan], and ICF International. (2013). *Tajikistan Demographic and Health Survey 2012*. Dushanbe, Tajikistan, and Calverton, MD: SA, MOH, and ICF International. Retrieved from http://dhsprogram.com/pubs/pdf/FR279/FR279.pdf.

Van Atta, D., Malvicini, P., Sigman, V., McNamara, P., Mueller, B., Liamzon, C., and Ludgate, P. (2015). Manual for the feed the future tajikistan household farm extension and advisory system. FAST project. Retrieved from http://meas.illinois.edu/wp-content/uploads/2015/04/FAST-2016-EAS-manual-Van-Atta-et-al.pdf.

This chapter is based on learnings from the United States Agency for International Development (USAID) and US Government Feed the Future FAST project. USAID is the leading American government agency building socioeconomic prosperity, together with the government and people of Tajikistan. The work was made possible by the generous support of the American people through USAID. The contents are the responsibility of the authors and do not necessarily reflect the views of the USAID or the US government.

17 Punjab Horticultural Postharvest Technology Centre

Building and Strengthening Linkages between Postharvest Professionals and the Emerging Horticulture Sector in India

B. V. C. Mahajan

CONTENTS

17.1 INTRODUCTION

The predominant rice-wheat system in the state of Punjab contributed significantly towards food security in India, but created several problems, such as depletion of the water table, poor soil health, aggravation of insect pests, overuse of pesticides, and environmental pollution, etc. Evidently, diversification from monocropping of wheat (*Triticum aestivum* L.) and rice (*Oryza sativa* L.) to horticultural crops for the conservation of both soil and water is essential. Growing of fruits, vegetables, and flowers have received an impetus in Punjab under the program of the National Horticulture Mission. In Punjab, fruits and vegetables occupy an area of 325,000 hectare with the production of 6.5 million metric tons (MT). The postharvest management (PHM) of fruits and vegetables is a highly specialized area and needs focused attention in terms of creation of infrastructure at farm and market levels for the overall development of horticulture.

Horticulture is a vibrant sector that is growing at 5% per annum in Punjab, thus making a dent in the state's dominant but stagnant food grain economy. The perishable nature of horticulture produce, with a significant amount of postharvest losses (PHL) annually, is not only a drain on the profitability of the farmers and other stakeholders but is affecting the growth of this high value segment of agriculture, which is contributing 10% to the state's Agriculture Gross Domestic Product (GDP) from only 4% gross cropped area.

With rising health consciousness of society in general, and growth of the middle class with its improved purchasing power, there is increasing demand for horticulture produce, which is placed higher in the food chain due to its health-promoting nutritional, medicinal, and other therapeutic properties. Further, it is reported that the rotting and decomposing of 1 MT of fruits and vegetables produces approx. 1.5 MT CO_2, which contributes to global warming, thus making PHM the main concern.

Given this background, the state's initiative to establish the Punjab Horticultural Postharvest Technology Centre (PHPTC) has become the need of the times. The envisioned intervention will surely save huge value by reducing PHL to benefit all the stakeholders in the chain, including farmers, thus making the horticulture sector more profitable and putting it on a faster growth trajectory.

17.2 BACKGROUND AND RATIONALE

Postharvest specialists from the University of California, Davis (UC Davis), assisted the government of Punjab in the design and launch of a new applied research and extension center in the discipline of postharvest horticulture in 1998. Dr. Michael Reid, Associate Dean, and Dr. Lisa Kitinoja, Office of International Programs, both of UC Davis, and Dr. James R. Gorny, Vice President, Davis Fresh Technologies participated as trainers and instructors in the first extension program held during the last week of August 1998 by the newly launched PHPTC in the city of Ludhiana, Punjab,

India. The design of the center and preparation for the launch of the PHPTC was funded by the United States Agency for International Development (USAID) through Chemonics International, under the Agricultural Commercialization and Enterprise (ACE) Project, managed by Mr. Keith Sunderlal of SCS Associates in New Delhi.

The State Government of the Punjab, in response to farmer concerns with associated loss of income, requested assistance in setting up a research and extension center with the purpose of reducing PHL. Lisa Kitinoja, as a specialist in postharvest extension education, was invited to Punjab to lead the process of strategic planning of the proposed center during January and February 1998. At that time, she interacted with over 60 stakeholders representing governmental agencies, university faculty and administrators, local, state, and national funding agencies, extension professionals, and clientele from throughout the horticultural sector in Punjab. These diverse groups shared the common interest in improving postharvest handling, packaging, cooling, storage, and transport of perishables, and were ready to take on the task of designing and funding the PHPTC.

It was further decided that the government of Punjab would fully fund the center through the Punjab Agricultural Marketing Board (known as the Mandi Board) on applied postharvest research and training activities. The PHPTC, located on the campus of Punjab Agricultural University (PAU), would house modern laboratory facilities, temperature control chambers, and a state of the art postharvest reference room (with internet access, videos, slide presentations, and written publications), manage demonstrations of recommended postharvest technologies in a variety of locales in Punjab, and offer a wide range of short-term training programs for small-scale and commercial clientele groups.

During the summer of 1998, James Gorny and Lisa Kitinoja were invited by Chemonics International to assist with the development of training materials oriented to small-scale produce marketers in India that were reproduced and marketed through the PHPTC. They prepared a 400-page book entitled *Postharvest Technology for Fruit and Vegetable Marketers: Economic Opportunities, Quality and Food Safety*, covering topics from planning production, harvesting, and packing through storage of fresh produce, small-scale processing, and marketing options for the Indian horticulturalists. Another manual on small-scale postharvest handling practices, written in 1995 by Kitinoja and Adel. A. Kader of the Department of Pomology at UC Davis, has been translated into Punjabi for dissemination to local horticultural farmers and small-scale marketers.

17.3   ABOUT THE PHPTC

The PHPTC was established in 1998 after signing a Memorandum of Understanding (MOU) between the Punjab government, through its agency Punjab Agri Export Corporation Ltd; PAU, Ludhiana; USAID, through ACE; and UC Davis. It was decided that

- State government would provide funds for the establishment of laboratories and operational costs
- PAU would provide scientific staff in the fields of horticulture, vegetable science, food science and technology, processing and food engineering, and chemistry, along with suitable building space
- UC Davis would provide technical support, such as training required for faculty, farmers, and traders

17.3.1   MISSION

In 2016–2017, the PHPTC was reorganized and relaunched with the mission "To be a leader in the postharvest management reducing losses, adding value, establishing cool chain to reduce global warming through collaborative efforts of the Government, the Universities, and the private sector to build and strengthen linkages between postharvest professionals and horticulture sector in India."

17.3.2 PHPTC Mandate

- To conduct training courses on postharvest handling for farmers, marketers, etc.
- To conduct applied research and develop appropriate postharvest technologies
- To assist the domestic horticulture industry to achieve international standards and to set a benchmark for quality produce in India
- To build and strengthen linkages between postharvest professionals and the horticultural sector in India
- To provide information on standards/specifications for export/domestic marketing of food
- To provide laboratory services for trade and industry (to estimate pesticides residue, quality parameters etc.)
- To provide reference room facilities for postharvest information

17.4 ACHIEVEMENTS

17.4.1 Research and Development

PHPTC conducted need-based research based on the feedback of farmers and industries, and developed the following protocols pertaining to postharvest handling, packaging, and storage of fruits and vegetables.

17.4.1.1 Enhancing Storage Life of Fruits

- **Storage of kinnow mandarin (*Citrus reticulata* Blanco):** The kinnow fruits harvested at optimum maturity and packed in ventilated corrugated fibreboard boxes can be stored for 45 days, with acceptable quality, at 5°C–6°C and 90%–95% RH.
- **Storage of pear (Hard pear and soft pear):** The "Patharnakh" pear (*Pyrus pyrifolia* [Burm. f.] Nakai) fruits can be stored in cold storage (0°C–1°C and 90%–95% RH) for 60 days with post-storage shelf life of two days at ambient temperature, and four to six days in the refrigerator. The "Punjab Beauty" pear (*Pyrus communis* L.) fruits can be stored in cold storage (0°C–1°C and 90%–95% RH) for 60 days with post-storage shelf life of one to two days at ambient temperature, and four days in the refrigerator.
- **Storage of ber (*Ziziphus mauritiana* Lam.):** The ber fruits of cultivar "Umran," harvested at color break stage, can be stored at 7.5°C ± 1°C and 90%–95% RH for two weeks with acceptable colour and quality.
- **Storage of plum (*Prunus salicina* Lindl.):** The "Satluj Purple" plum fruits, harvested at color break stage followed by postharvest treatment of a calcium nitrate (2%) solution for five minutes, can be stored for four weeks in cold storage (0°C–1°C and 90%–95% RH) with post-storage shelf life of two days at ambient temperature.
- **Storage of grapes (*Vitis vinifera* L.):** The "Flame Seedless" grapes, harvested at optimum maturity with firm berries having light purple color, packed in ventilated corrugated fiberboard boxes (4 kg) lined with polythene film containing one sheet of grape guard, can be stored with acceptable quality for 45–50 days at 0°C–2°C and 90%–95% RH.
- **Chlorpropham-assisted storage for maintaining low reducing sugars in ware potato (*Solanum tuberosum* L.):** Potato cultivars "Kufri Chandermukhi," "Kufri Jyoti," and "Kufri Chipsona-1" can be stored successfully for five months at 10 ± 1°C and 90%–95% RH, with two consecutive foggings of Chlorpropham (CIPC) at the rate of 40 ml/t. The first fogging is given at the initiation of sprouting, and the second 60 days after the first fogging. The stored potato maintain low reducing sugars (<0.25%) and are suitable for preparation as chips and for other culinary purposes.

17.4.1.2 Ripening Technology for Fruits

- **Ripening of banana (*Musa paradisiaca* L.) and mango (*Mangifera indica* L.) with ethylene gas:** The banana and mango fruits harvested at green mature stage can be successfully ripened in four days by exposing them to ethylene gas (100 ppm) for 24 h in a ripening chamber maintained at 16°C–18°C and 90%–95% RH for banana and 20°C–22°C and 90%–95% RH for mango.
- **Ripening of banana with ethrel treatment:** The banana fruits harvested at the green mature stage can be successfully ripened in four days by dipping in a solution of ethephon at 500 ppm (1.25 ml per litre of water) for 2–3 min. The fruits should be air-dried and kept at 16°C–18°C and 90%–95% RH. The fruits attain uniform color and excellent quality, with a shelf life of four days at 16°C–18°C and two days at 30°C–32°C.
- **Ripening of pear with ethylene gas and ethrel treatments:** Ethephon at 1000 ppm as postharvest dip or exposure of fruits to ethylene gas (100 ppm) for 24 h in a ripening chamber, followed by storing the fruit at 20°C for ripening, proved to be the most effective treatments for improving ripening and maintaining the marketable quality of "Patharnakh" pears. The fruits took eight days to ripen at 20°C, while four days at ambient conditions.
- **Ripening of tomato (*Solanum lycopersicon* L.):** Winter season tomatoes harvested at color break stage, free from bruises and diseases, packed in plastic crates lined with newspaper, can be ripened in seven to 10 days in ventilated polyhouse conditions or a ripening chamber (20°C temperature and 85%–90% RH). The fruit attains uniform color and quality during ripening. However, the sorting of ripened fruits should be carried out at regular intervals to avoid further PHL.

17.4.1.3 Shrink Wrapping of Fruits and Vegetables for Retail Marketing (Table 17.1)

The technologies developed for shrink wrapping of fruits and vegetables for retail marketing have been summarized in Table 17.1.

17.4.1.4 Drying Technology for Vegetables

- **Drying of Kasuri methi (*Trigonella corniculata* L.):** Methi can be dried successfully in a mechanical drier with a drying time of 9 h (2 h at 70°C, 3 h at 60°C and 4 h at 50°C).
- **Drying of bitter gourd (*Momordica charantia* L.):** Bitter gourd slices of 1–2 cm thickness can be optimally dried with acceptable quality at 65°C/2 h; 55°C/8 h, followed by equilibration at ambient conditions (35°C–40°C).

TABLE 17.1
Observed Shelf Life of Shrink Wrapped Fruits and Vegetables

Commodity	Packaging Film	Shelf Life under Ambient Conditions
Kinnow mandarin	Shrink	15 days
Daisy mandarin	Shrink	15 days
W. Murcott mandarin	Shrink	10 days
Peach	Shrink/Cling	4 days 9 days (under supermarket conditions)
Capsicum	Shrink	10 days
Tomato	Shrink/Cling	6 days
Cabbage	Shrink/Cling	15 days
Brinjal	Shrink/Cling	7 days

17.4.1.5 Development of Low Cost Onion (*Allium cepa* L.) Storage Structures

Mechanically ventilated mild steel onion storage structures of 1.25 MT capacity, comprising a top cuboid portion of 1.2 × 1.2 × 1.2 m and a bottom triangular prism of 1.2 × 0.6 × 1.35 m with a 35° angle of inclination, coupled with mechanical ventilation of air in the range of 0.75–1.0 ms^{-1}, can safely store this commodity up to five months with acceptable quality.

17.4.1.6 Design of a Plastic Crate and Corrugated Fiber Board Box (CFB) for Packaging of Tomatoes

A plastic crate of internal size 440 × 285 × 135 mm can hold about 10 kg of tomatoes, while corrugated fiberboard box of internal size 325 × 210 × 180 mm (3–5 ply) can carry 5 kg of tomatoes for distant and domestic marketing with minimum PHL.

17.4.2 ESTABLISHMENT OF LABORATORIES

The PHPTC has established three state of the art laboratories, the Postharvest Laboratory, the Quality Control Laboratory for Processed Foods, and the Grain Moisture Calibration Laboratory. The setup of these laboratories is as follows.

17.4.2.1 Postharvest Laboratory

The Postharvest laboratory is fully equipped with walk-in cold storage, precooler, ripening chamber, hydrocooler, package testing machines, shrink packaging machine, cling wrapping machine, Hunter color meter, moisture analyzer refractometers, penetrometers, etc.

17.4.2.2 Quality Control and General Analysis Laboratory

The Quality Control and General Analysis laboratory contains an Atomic Absorption Spectrophotometer (AAS), Gas Chromatography Mass Spectrophotometer (GC-MS), microwave digestion system, and many other supporting instruments.

17.4.2.3 Moisture Meter Calibration Laboratory

The PHPTC has established a moisture meter calibration and repair laboratory, which regularly provides services to the Punjab Mandi Board for calibration of moisture meters before procurement of wheat and paddy.

17.4.3 CAPACITY BUILDING THROUGH TRAINING PROGRAMMES

The PHPTC has prepared need-based training modules of one day, five day, and three week duration on PHM, value addition, and marketing of fruits and vegetables for farmers, traders, processors, exporters, extension specialists, and scientists. The training programs consist of hands-on training, demonstrations, and specialized activities. The training programs were designed to familiarize farmers and other professionals with various techniques used during supply chain management of fruits and vegetables, such as harvesting, grading, washing/waxing, packaging, transportation and storage, etc. The center has organized more than 350 training courses (Figures 17.1 and 17.2). The PHPTC has also conducted training in collaboration with international agencies like UC Davis and Programma Uitzending Managers (PUM), Holland.

17.4.4 REFERENCE LIBRARY

PHPTC has established its own library, where literature pertaining to the PHM of horticultural crops has been purchased or downloaded from Internet sites. The literature is placed in the folders for use by readers in the form of books, manuals, and journals published in India and abroad.

FIGURE 17.1 PHPTC scientists and other staff.

FIGURE 17.2 Training programs conducted by PHPTC scientists.

17.4.5 LINKAGES

The scientists of the PHPTC have received funding for research and training programs for the establishment of a state of the art quality control laboratory, postharvest laboratory, and training facilities from various agencies like Ministry of Food Processing and Industries, New Delhi; National Horticulture Board, Gurgaon; and National Horticulture Mission, New Delhi. In addition, many private companies, such as United Phosphorus Ltd (UPL), Container Corporation (CONCOR), G.N. Packaging Ltd, Nipro Fresh, and RajHans Fertilizer Ltd, have collaborated with PHPTC to test postharvest chemicals, wax coatings, packaging materials, etc. The scientists of the PHPTC are invited, as are source persons, to various institutes, such as Central Institute for Postharvest Engineering and Technology, Ludhiana; Guru Angad Dev Veterinary University, Ludhiana, and many other reputed universities in India, for delivering specialized lectures.

17.5   ROLE OF PHPTC IN THE ESTABLISHMENT OF POSTHARVEST INFRASTRUCTURE IN PUNJAB STATE

PHPTC is working in close collaboration with Punjab Mandi Board, Punjab State Department of Horticulture, and Punjab Agro Industrial Corporation to provide technical input for the establishment of modern postharvest and cold chain infrastructure. The scientists also provided technical know-how to farmers for the export of kinnow fruits from the Punjab to Russia, Ukraine, Thailand, etc. The details are as follows.

17.5.1 POSTHARVEST INFRASTRUCTURE CREATED BY PUNJAB MANDI BOARD

PHPTC provided technical consultancy to Punjab Mandi Board for the establishment of the state of the art postharvest infrastructure in 13 fruit and vegetable markets of Punjab, comprising 16 multicommodity cold storages of 800 MT capacity and 52 ripening chambers of 520 MT capacity (Table 17.2).

TABLE 17.2
Cold Stores and Ripening Chambers Established by Punjab Mandi Board at Different Agriculture Produce Market Committee (APMC) of Punjab State, India

Serial No.	Name of APMC	Ripening Chamber (Capacity 10 MT Each Chamber)	Cold Room (Capacity 50 MT Each Chamber)
1	Ludhiana	6	5
2	Sangrur	2	1
3	Moga	4	1
4	Jalandhar	8	2
5	Nawanshahar	4	1
6	Batala	2	1
7	Ferozepur	2	1
8	Abohar	2	1
9	Patiala	2	1
10	Hoshiarpur	6	2
11	Bathinda	6	2
12	Kapurthala	2	1
13	Phagwara	2	1
	Total	**48**	**20**

17.5.2 Postharvest Infrastructure Created under National Horticulture Mission (NHM)

PHPTC provided technical consultancy to the center of excellence for fruits, vegetables, and potato for the establishment of precoolers, cold storage, and packaging machinery. Technical knowledge was also provided to various private entrepreneurs for the establishment of cold storage, ripening chambers, and integrated packhouses who received financial assistance under the National Horticulture Mission program. The detail is as follows.

17.5.2.1 Cold Storages and Ripening Chambers Established under NHM

In Punjab, there are about 560 cold stores having a capacity of about 1.96 million t, out of which 139 cold stores of 563,000 t capacities and 92 ripening chambers with 1255 t capacities are established by various private entrepreneurs with the financial assistance from NHM (Table 17.2).

17.5.2.2 Farm-Level Packhouses

Under NHM, approximately 472 packhouses have been set up in the state for fruits and vegetables. The packhouse is the basic need of the horticultural sector for collection, grading, and sorting of produce.

17.5.3 Postharvest Infrastructure Created by Punjab Agro Industry Corporation (PAIC)

PHPTC provided technical consultancy to Punjab Agro Industry Corporation for the establishment of washing, waxing, and grading lines for kinnow mandarin and postharvest infrastructure at "Mega Food Park." The scientists also provided technical know-how to farmers for the export of kinnow mandarins, through PAIC, from the Punjab to Russia, Ukraine, Thailand, and other countries.

17.5.3.1 Mechanical Washing, Waxing, and Grading Lines

At present, there are five mechanical washing, grading, and waxing plants established by PAIC. These washing and waxing plants are established at Chhauni Kalan and Kang Mai in Hoshiarpur district; Abohar, Jattan Tahliwala in Ferozepur district; and Badal in Muktsar district. Besides, many progressive growers (more than 200 in number) have also installed such facilities privately at their own farms.

17.5.3.2 Processing Plants

The Punjab Government has established two pilot-scale processing units (20 t per hour capacity) for processing of kinnow juice and other fruits and vegetables at Hoshiarpur and Abohar.

17.5.3.3 High Tech Packhouses

Five packhouses with cold storage and grading/sorting lines have been established by Punjab Agri Export Corporation (PAGREXCO) at Mushkabad (Ludhiana), Saholi and Lalgarh (Patiala), Kangmai (Hoshiarpur), and Babri (Gurdaspur).

17.5.4 Adoption of Technologies by Farmers and Entrepreneurs

The various technologies toward harvesting, grading, packaging, and storage of fruits and vegetables developed by PHPTC have been commercially adopted by farmers and entrepreneurs (Figure 17.3).

FIGURE 17.3   Adoption of PHM technologies by farmers and entrepreneurs.

17.6   PUBLICATIONS

Scientists of the center have published research papers, popular articles, and research reports, as well as a training manual and a compendium of lectures delivered in refresher courses.

Research papers	>100
Popular articles	>100
Training manual	2 (English and Punjabi)
Refresher courses	4
Newsletter	Quarterly

17.7   PHPTC WEBSITE

PHPTC has developed its own interactive website (www.phptc.org) for the benefit of farmers, traders, processors, students, scientists, and extension specialists. The website is regularly loaded with recent research and training activities and literature, etc.

17.8 FUTURE THRUSTS: VISION

The vision for the future involves reducing PHL and ensuring a higher profit to the farmers and other stakeholders through improvement in the PHM, supply chain, and market linkages of horticultural produce.

The horticulture sector in Punjab is swiftly surging ahead with an increasing share of horticulture in the state's Agriculture GDP from 6.11% in 2004–2005 to 10.30% in 2015–2016. Though many factors contribute to such growth, the perishable nature of horticulture produce and the "postharvest management factor" for this category of produce assumes increased importance. Horticulture production in Punjab has increased from 3.4 million t to 6 million t in the last decade, wherein the PHL of horticulture produce is assessed to be around 20%. It is a heavy drain on the state's economy, as well as the profitability of farmers who take risks for diversification. PHL and falling prices during glut periods can be converted into profitable value added products using efficient PHM protocols and cold chain infrastructure facilities. This in turn improves shelf life and availability of stored/value added or fresh quality horticulture produce to the consumers in distant markets when and wherever there is a high demand.

Rising health consciousness and awareness against pesticide residues in the local and distant domestic market, as well as increasing export potential with strict quality specifications, require standardized PHM protocols and accredited laboratory testing.

With this background, it is planned that PHPTC will contribute significantly to developing and strengthening the horticulture sector, with special emphasis on postharvest and cold chain infrastructure in the region. To achieve this, various activities of the center will be strengthened through following capacity building programs like the following.

17.8.1 TRAINING AND HUMAN RESOURCE DEVELOPMENT

PHPTC will function with a professional approach in providing training to the entrepreneurs, cold store owners, self-help groups, progressive farmer's organizations, agro-marketing cooperatives, etc., on safe handling, storage, and value addition of horticultural produce.

17.8.2 DEVELOPMENT OF PROTOCOLS FOR DOMESTIC AND EXPORT MARKETS

Fruits and vegetables are perishable commodities and require tender care after harvesting in order to avoid PHL. Marketing of fresh fruits and vegetables has become more challenging day by day due to increasing global competition for quality produce. Therefore, there is need to develop protocols, such as standardization of maturity indices, grading, eco-friendly packaging, postharvest treatments, edible coatings, storage, and safe ripening techniques for domestic, distant, and export markets. The PHPTC will invite private companies (dealing with molecules, chemicals, coatings, packaging materials, etc., used for shelf life enhancement) to get their products tested at reasonable rates, for the benefit of the horticultural industry of the state as well as India.

Based on the feedback of the industry and farmers, the center will undertake problem-oriented research studies on PHM of horticulture crops on a pilot scale for optimization of temperature, treatments, packaging, etc., for maintaining quality and reducing PHL.

The problems related to harvesting, precooling, packaging, storage, transport, and destination handling will be identified through active interaction with different organizations, such as Citrus Estates, Pear Estates, Litchi Estates, Horticulture Department, Punjab Mandi Board, Punjab Agro Industry Corporation, etc.

Concerted efforts will be made to reduce the PHL of fruits and vegetables by generating eco-friendly and suitable postharvest technologies, as well as dissemination of these technologies through extensive hands-on training programs, seminars, field days, etc., for farmers, storage operators, marketers, and other stakeholders. Focused efforts to use the cold chain as a weapon to manage glut during peak season and promoting entrepreneurship will generate more value to the horticulture sector.

17.9 CONCLUSION

Horticulture has been identified as an important sector for diversification of agriculture to ensure faster and sustainable agriculture development in Punjab. However, because of the lack of awareness about postharvest handling practices and non-availability of adequate postharvest infrastructure, a huge quantity of valuable produce goes to waste. Keeping in view the challenges of PHL and global competition for the export of fruits and vegetables, there is an urgent need to focus on formulating research and training programs in the direction of postharvest handling, packaging, and storage of fresh horticultural produce. PHPTC is working as a nodal agency to guide farmers, entrepreneurs, and government agencies for the establishment of postharvest and cold chain infrastructure at farm and market levels. The various PHM technologies developed by PHPTC, such as the ripening of banana, packaging and storage of kinnow, grape, plum, peach, pear, ber, potato, tomato, capsicum, cabbage, etc. have been approved by PAU, Ludhiana for commercial adoption by farmers and aspiring entrepreneurs. The Punjab State Agriculture Marketing Board started a "farmers market" also known as *Apni mandi*, with a view to giving a boost to the small farmers around cities so as to provide direct access to the consumers by eliminating the middleman. PHPTC has been providing hands-on training to *Apni Mandi* farmers and other stakeholders of Punjab state since its inception. PHPTC would endeavor to generate farmer friendly, as well as export-oriented, technologies for fresh horticultural produce. The PHPTC as described in this chapter can serve as a successful model extension approach for reducing PHL in other countries.

ACKNOWLEDGMENTS

The significant contributions of former Directors of PHPTC, Dr. A.S. Dhatt and Dr. B.S. Ghuman, and other individuals involved in the establishment of PHPTC and designing of training and research programs, is duly acknowledged.

The author would also like to express special appreciation and thanks to Dr. Lisa Kitinoja, International Postharvest Consultant, Oregon, USA for reviewing, editing and providing valuable suggestions in compiling this chapter.

18 Monitoring and Evaluation Practices for Improving Postharvest Extension Projects

Lisa Kitinoja and Devon Zagory

CONTENTS

18.1 OVERVIEW OF MONITORING AND EVALUATION PRACTICES FOR POSTHARVEST PROJECTS AND PROGRAMS

Monitoring and Evaluation (M&E) during postharvest training programs and projects is a very important part of any postharvest project. Often M&E is missing or is done in a perfunctory manner when financial resources are limited. Evaluation specialists recommend including a minimum of 5% in the budget for M&E purposes, although this is rarely done in practice.

It is important to monitor and evaluate postharvest projects and programs for three major reasons.

- Accountability. M&E may be required by donors or project managers.
- Making program improvements. This is known as formative evaluation, and is used to assess progress and make changes and adjustments while the program or project is being implemented.
- For future project planning/proposal development. This is known as summative evaluation, and is typically done at the close of a project or program with the aim of assessing the success in achieving objectives.

Typically, M&E measures the changes that can be attributed to the program or project, so there is a need for a baseline measurement of "indicators" to characterize the current situation.

Examples of important changes in postharvest projects or programs include:

- Developing new interests or aspirations regarding reducing food losses
- Gaining new knowledge of improved practices or technologies
- Learning new skills to use improved postharvest practices or technologies
- Adopting new postharvest practices
- Investing in new postharvest technologies
- Accessing new sources of income or increasing profits
- Measuring changes in food losses

Changes can be positive or negative, intended or unintended. Sometimes a change can lead to a negative effect, such as when workers (generally women) who process foods manually are replaced by a machine (generally operated by a man).

A theory of action or logic model can help to attribute any measured changes to the program's activities (using **if-then** logic). In the case of the 2016 Postharvest Education Foundation (PEF) Training of Trainers (ToT) program, we can compare the results of a Training Needs Assessment survey of e-learners conducted at the end of the program (posttest) with the one they completed at the beginning (pretest). Example: **If** e-learners gain new knowledge and skills, and apply their knowledge/skills to provide postharvest training programs for their community, **then** local trainees will gain new knowledge/skills and **then** they may adopt better postharvest practices. **If** better postharvest practices are adopted, **then** food losses will be reduced, and participants will increase their incomes.

The monitoring system should include three interrelated practices or procedures: financial and implementation monitoring, beneficiary assessments, and diagnostic studies (Coote et al. 2002).

1. Implementation monitoring looks at whether the project interventions are being delivered on time and within budget.
2. Beneficiary assessments determine whether the interventions are considered relevant by participants, and whether any promoted technologies are or are not being adopted.
3. Diagnostic studies may be undertaken to try and adjust the project activities to address any problems (such as if the technologies are being rejected or target groups are not reacting as expected).

Evaluation systems are distinct from monitoring systems, and they may be carried out by different people for different purposes, but project evaluation typically makes use of the information gathered during monitoring (Coote et al. 2002). For this reason, a high-quality monitoring system will make the project evaluation (formative and/or summative) processes quicker and easier.

18.2   SETTING MEASURABLE OBJECTIVES

Setting measurable objectives is a key component of successful projects. Following a structured process can facilitate the setting of those objectives. Such a process would first clearly articulate the project's mission and goals, followed by a statement of objectives leading to a theory of action. The results of those actions are what are subject to the measurement of the objectives. Typically, a postharvest project or program will have the following steps spelled out in the proposal and planning stages:

1. *Mission*: The purpose and grand plan for the future. Example: No more food insecurity on Earth!

2. *Goals*: The long-term achievements of the postharvest project or program. Example: To reduce food losses and thereby improve incomes for participants in the program so they can be food secure and healthy.

3. *Objectives*: A measurable set of steps or actions that will help to reach the stated goals. Example: To provide postharvest training of farmers in order to help them to learn better postharvest handling practices for their crops leading to reduced food losses and increased incomes.

4. *Theory of Action* (also known as a Results Chain or Program Logic): The theory of action is a chain of activities and/or events that can take us from where we are now to where we want to be, and therefore achieve the project objectives. A theory of action uses an "If-Then" logic statement. Example: **If** the training provided by the program reaches people who learn enough to **then** make practical changes and adopt better postharvest technologies, **then** the current levels of high postharvest food losses will be reduced, and earnings increased by selling more of their crops.

18.2.1   MEASURABLE OBJECTIVES INCLUDE THREE COMPONENTS

- *Indicators* are parameters that can be counted (such as numbers of sites or people to be reached, numbers of postharvest topics or technologies to be demonstrated).
- *Targets* are the goals of having a certain number of people participating in a program or making changes in practices, or measuring a specific desired change in % losses.
- *Timeline*, such as the number of participants per week or per year.

As an example, when the project evaluation of the European Community-funded Strengthening Livelihoods and Food Security through Reduction of Grain Post-harvest and Storage Losses in Southern Somalia (IPHASIS) was conducted, the first step was to evaluate the overall project objective, stated as "Food security of the target 16,700 farmer households (40% women) is enhanced by reducing grain storage losses through improved post-harvest and storage" (i.e., hermetic silos and dam liners for pit storage) in the course of 3 years (LTS Africa 2012). The indicators and targets in this statement include the number of farmer households, % of women and number of technologies to be promoted. The timeline was the full length of the project period.

The findings determined that the objective was relevant, and the design of the project was originally sound as regards targeting the real needs and problems of the right beneficiaries. The project addressed capacity building to reduce losses using two improved postharvest technologies, and "sought to improve existing storage technologies as opposed to introducing completely new technologies which the beneficiaries might have taken a longer time to accept."

- The introduction of silos was a simple upgrade for the traditional drum that most people used for maize storage. The making of the hermetically sealed metal silos using metal inert gas (MIG) welding did not prove difficult for the Somali artisans as they already had soldering skills.
- The dam liners, substituted for the maize stalks and cattail (*Typha latifolia*) straw that lines the traditional pit, were easily accepted after demonstrations showed that the use of the dam liners was better than the traditional methods because the problems of moisture and attacks by pests were effectively addressed.

18.3 THE INDICATORS USED TO MEASURE CHANGES IN POSTHARVEST LOSSES

There are many different levels of indicators that can be used to assess or measure changes. The higher the level of the indicator (as illustrated in Figure 18.1), the easier it can be for the evaluation to demonstrate the direct results (outputs, outcomes, and/or impacts) of the project activities and inputs.

- Outputs are under the direct control of the project or program.
- Outcomes are short-term or medium-term effects.
- Impacts are long-term effects, which may take years to develop.

Indicators must be measurable and can be quantitative or qualitative. Indicators are often named, with specific targets stated, in the program/project objectives. When objectives are well-written, it is easier to identify appropriate indicators and targets.

Targets for each indicator (the setting of a specific measurable amount of change) can be based upon baseline data, expert opinions, stakeholder expectations, historical trends, and/or accomplishments of similar programs.

For example, when PEF offers the postharvest ToT program to 100 participants, our first target would be to document the completion of the program by at least 75 persons in 1 year (indicator, target, and timeline one). Of these ~75 participants, our next target is to document their increase in postharvest knowledge and skills by 20% to 30% compared to their initial level (indicator, target, and timeline two). Within the next year, our target is for 80% of those who complete the training to offer a postharvest training program to an audience of 30 persons in their community (indicator, target, and timeline three).

Bennett's Hierarchy of Evidence is one example of a logical framework where inputs and activities create outputs that can lead to outcomes and impacts (Bennett 1975). Having clear indicators for each level will help to establish a plausible link to explain any measured changes.

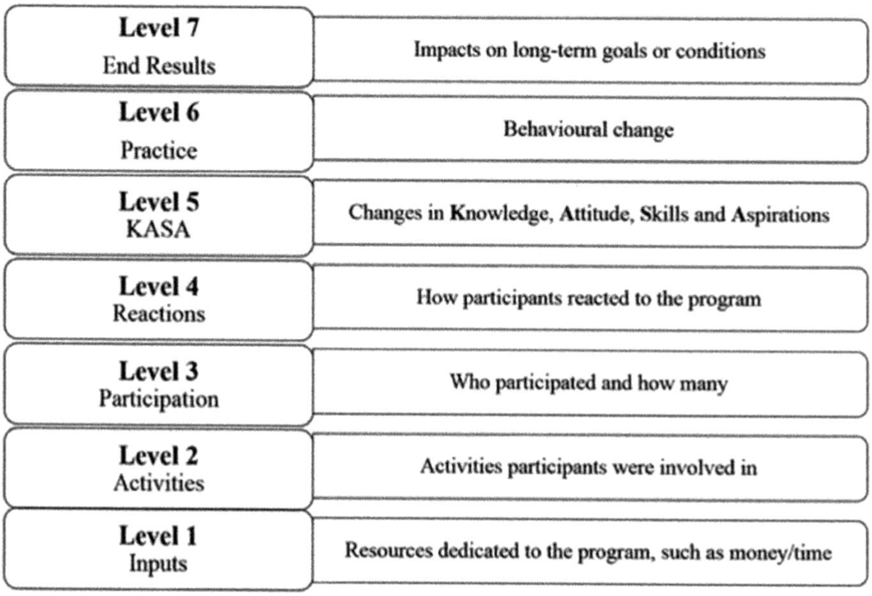

Level 7 End Results	Impacts on long-term goals or conditions
Level 6 Practice	Behavioural change
Level 5 KASA	Changes in Knowledge, Attitude, Skills and Aspirations
Level 4 Reactions	How participants reacted to the program
Level 3 Participation	Who participated and how many
Level 2 Activities	Activities participants were involved in
Level 1 Inputs	Resources dedicated to the program, such as money/time

FIGURE 18.1 The seven (7) levels of Bennett's hierarchy of evidence. (Modified from Bennett, C., *J. Extension*, 13, 7–12, 1975.)

- Level 1: Inputs
- Level 2: Activities
- Level 3: Participation (outputs)
- Level 4: Reactions of participants (short-term outcomes)
- Level 5: Changes in participant knowledge/attitudes/skills/aspirations (short-term outcomes)
- Level 6: Changes in participant behaviors or adoption of new practices (medium-term outcomes)
- Level 7: End results (long-term impacts)

Objectives can be written in reference to any of the seven levels of Bennett's Hierarchy. A complex objective will include indicators that are related to two or more levels. In the case of the PEF e-learning program, we want to determine the short term outcomes (changes in interest, knowledge, skill and/or experience level) of our participants.

18.4 EVALUATION DESIGN

An evaluation design is the overall plan for making comparisons. There are several types of designs used to plan evaluations and many free to access reference books and manuals that can be used to develop evaluation plans and designs (Donaldson 2014; IFRC 2011; Austrian Development Agency 2009; NSF 2002). In addition, there are two classic "Evaluation" textbooks still in wide use that provide full details (Patton 1991; Rossi and Freeman 1982; Rossi et al 2003). Some important designs are:

- Experimental design, which requires random selection and assignment, and large sample sizes, so may be difficult to achieve
- Quasi-experimental research design, which shares similarities with the traditional experimental design or randomized controlled trial, but it specifically lacks the element of random assignment to treatment or control
- Comparison to a baseline (if the indicators are expected to change over time)
- Comparison to a control group (if the program participants will show changes that are more or different than those of a similar group that did not participate)

Including stakeholders in the M&E planning and implementation process will improve the chances that the evaluation results are utilized for decision-making and future planning.

18.5 DATA COLLECTION METHODS

Collecting data for M&E purposes is similar to collecting data for a research study.
Types of data and typical data collection methods include:

- Quantitative data (statistics, counts, numbers, costs, etc.) can be collected via making measurements, conducting formal surveys, and analyzing secondary databases.
- Qualitative data (on perceptions, beliefs, ideas, aspirations, behaviors, etc.) can be collected via observations, interviews, rapid rural appraisals, Commodity Systems Assessment Methodology (CSAM), and focus groups.

Questionnaires or surveys can be very simple and to the point, if the indicators are clearly specified. An example of an evaluation data collection plan is provided in Table 18.1 from a Postharvest Training and Services (PTSC) pilot project (Kitinoja and Barrett 2015). Data were collected using a variety of methods (surveys, face-to-face interviews, site visits) as shown in Table 18.1, where X indicates which method(s) were used for which target groups.

TABLE 18.1

Data Collection Plan and Implementation

Data Collection Plan and Implementation

Target Groups	Size of Target Groups	Email Surveys	Face-to-Face Interviews	Phone Interviews	Site Visits for Observations	Response Rate
Training of trainers (ToT) participants	36 in total	X				92%
Local trainees (farmers and food processors)	50 people, random cluster sample selected from a population of 500 trainees		X		X	100%
Postharvest trainers	14 in total	X		X		100%
PTSC administrators and managers	7 in total		X		X	100%

- **Which of the ten postharvest demonstrations offered by the PTSC do you consider most useful? (mark in column A with X), and estimate what they would cost (column B).**
- **Do you know of any trainees who have adopted changes in these postharvest practices or invested in new technologies? (mark in column C and D with X).**

Demonstration topic	A) Indicate the most useful demos (choose as many as you like)	B) Estimate the costs for mounting these demos if you want to do so on your own	C) Practice changes? * See below for follow-up	D) Investments? # See below for follow-up
Use of Shade				
Gentle handling				
Use of maturity indices (color charts, sizes)				
Sorting/grading				
Hand washing				
Zero energy cool chamber (ZECC)				
CoolBot™ equipped cold room				
Solar drying practices				
Making jams and preserves				

If Yes * (Column C), please describe any adoption of practice changes:
If Yes # (Column D), please describe the investments made by trainees:

FIGURE 18.2   Sample questionnaire from TOPS funded evaluation of PTSC project, Arusha.

The following example of a section of a questionnaire is from the Technical and Operational Performance Support (TOPS) (United States Agency for International Development [USAID]/Food for Peace program) funded evaluation of the Tanzania PTSC project (Figure 18.2; Kitinoja and Barrett 2015). The project implemented 10 postharvest demonstrations in 2012 and returned to evaluate their adoption in 2014–2015 by interviewing a group of postharvest trainers who had used the demonstrations to train local farmers, traders, and processors.

18.6 COST/BENEFIT ANALYSES

The costs and benefits of any change in practice or technology being promoted by a postharvest project or program need to be monitored and evaluated. PEF provides a worksheet for anyone who is assessing the relative costs and benefits of a postharvest practice or technology.

How do you estimate the changes in % losses? Traditional (current) practices and technologies have been documented to lead to high postharvest losses (PHL), in some cases as high as 50% to 80%. The main causes of high losses for perishable crops include poor quality containers, rough handling, high temperatures during the postharvest period, and delays in marketing (Kitinoja and Al Hassan 2012). The main causes for high losses of grains, dry beans, and other staple crops include spillage during harvesting, poor drying practices, and inadequate storage containers or structures (lack of protection from pests, heat, rain, or moisture). Improved practices and technologies can greatly reduce these losses. Using plastic crates can reduce damage and losses of fresh produce to less than 5% (Kitinoja 2013). Cool storage of perishables and proper dry storage for staples can reduce losses by half or even more, depending on the crop.

18.6.1 DETERMINING RELATIVE COSTS AND EXPECTED BENEFITS

PEF has developed a simple worksheet to compare any two practices or technologies. Costs and benefits are estimations, and the idea is to provide just one example of how farmers or marketers can improve their incomes by using a technology or handling practice that you are teaching them about. By comparing the current or traditional practice to a new technology or practice that you want to demonstrate for reducing food losses, you can use actual local market prices and determine whether it will be cost effective or not. You should consider only those aspects (equipment, supplies, labor, etc.) that will be different, and ignore any costs that stay the same. The amount available for sale will depend upon your estimations of % PHL.

Choose one crop and one postharvest practice change and consider all the attendant changes in costs and expected benefits. To learn how to use the worksheet, choose a simple practice change or a new technology, not something that is very complex (a complex change would combine many new practices and new technologies). The trick to using this simplified cost-benefit ratio (C/B) determination method is to start with a simple unit (1000 kg of produce harvested) and ignore any cost that is the same for both practices. We are interested only in a relative cost difference (does one practice cost more to use than the other?) and the relative benefits (does one practice lead to higher earnings than does the other?). For example, it costs more to use plastic crates, but the returns (market value) are also higher than for the traditional containers because the crates do a much better job of protecting produce than using sacks or baskets during transport and storage.

You may or may not have recurring costs to consider. In our example, we can use the plastic crates many times, but we need to purchase new liners (heavy paper or fiberboard) with each use. This helps to protect the crop from abrasions and to minimize contamination issues.

Sometimes the local market price will depend on quality or size (Grade 1, Grade 2, etc.), or the time of the year (glut time when there is oversupply or lean season when there is scarcity in the markets) so you may have more than one price to consider. Your new technology or handling practice may give you more Grade 1 than does the traditional or current practice, or it may provide more crop during the off-season when market prices are higher.

To complete the C/B worksheet for your postharvest technology, remember to think simple and small. For training purposes, try to use a simple example, one without any recurring costs. In each case, always begin with 1000 kg of the harvested crop—this makes the math easy to do. If you have more or less crop to work with, then after the worksheet is complete you can adjust the amount. For example, if you have 2000 kg at harvest, you can multiply the results by two. If you have only 500 kg at harvest, you can divide the results by two.

Cost/ Benefit Work sheet		
Description	**Traditional Practice**	**New Practice**
Costs		
Relative cost		
Recurring costs		
Expected benefits		
% losses (estimation)		
Amount available for sale		
Value/kg		
Total market value		
Market value minus relative costs		
Relative profit		
Return on Investment (ROI)		

* How many loads does it require to reach 100% Return on Investment?

FIGURE 18.3   Sample of simplified cost/benefit worksheet used by postharvest trainers.

If the farmers or food processors are very small-scale, it may be useful to complete the worksheet for your demonstration using only 100 kg for the load size (rather than 1000 kg).

The simplified cost-benefit worksheet used by postharvest trainers (Figure 18.3) was first developed by Lisa Kitinoja in the 1990s for extension education projects in Egypt, Lebanon, and India. Literate farmers and food processing cooperative leaders had no trouble understanding how to use the worksheet and explain their assumptions and calculations to fellow workshop participants. Calculations are greatly simplified by using 100 or 1000 kg as a base unit and ignoring any costs that are the same for both practices.

18.6.2   CALCULATING THE RETURN ON INVESTMENT

The return on investment (ROI) that you determine for the crop and technology combination may be negative or positive, and could be immediate or require several or many uses before you reach the break-even point. For example, even though the cost for plastic crates is very high, the new practice of using plastic crates instead of sacks may be immediately profitable with the first load because of greatly reduced % losses and a higher market price per kg. A negative ROI would indicate that the technology is not a good choice for that crop or location at that time of year.

18.7   CASE STUDY OF A POSTHARVEST PROJECT EVALUATION

Evaluation of the five key components of the Horticulture Innovation Lab funded PTSC pilot project (established in Arusha, Tanzania in 2012 and operating until funding ended in October 2013) from October 2014–April 2015 revealed that the two training-oriented components were fully implemented and resulted in outcomes that far exceeded their targets. The project was evaluated by World Food Logistics Organization (WFLO) for TOPS/Save the Children (Kitinoja 2015; Kitinoja and Barrett 2015).

18.7.1 Introduction

The project was designed to tackle the complex issues of reducing food losses and improving food security. While more than a dozen past international projects have identified appropriate postharvest technologies, and recommended a variety of training, capacity building, and small-scale infrastructure developments (Kitinoja 2010), this was the first project to integrate all of this technology and information and offer a locally based solution. This unique pilot project for smallholder farmers and food handlers in Arusha, Tanzania, known as the PTSC, combines a wide variety of training programs, adaptive research, and demonstrations of postharvest practices and services aimed at reducing losses and increasing shelf life. Via a postharvest retail shop, it provides on-site ready access to the tools and supplies people need in order to reduce PHL and improve market access and incomes for the smallholder farmers, women farmers, and village level processors who are affiliated with established cooperatives and farmers associations near Arusha in the northern zone of Tanzania.

18.7.2 Theory of Action

The project's Theory of Action can be described as follows:

Training a cadre of young postharvest horticultural trainers in Sub-Saharan Africa (SSA) as "postharvest specialists" will lead to their providing postharvest training programs for local farmers, traders, food processors, and marketers in their home countries. These trainees will gain new skills and the knowledge needed to make small investments and adopt new postharvest practices. These new postharvest practices will lead to reduced food losses, improved nutrition and/or increased incomes for their families. At the same time, providing a local venue for postharvest training, services, and retail goods and supplies in Tanzania will lead to local population involvement in horticultural activities by having a place to go for training, and to purchase the needed tools, packages, and other supplies they will then use for improving postharvest handling, storage, food processing, and/or marketing. These locally trained people will achieve their own goals of decreasing food losses, improving food security, and improving their livelihoods.

The PTSC model includes five components, some of which were designed to provide income in order to make the PTSC financially self-sustaining (Table 18.2).

TABLE 18.2
Project Objectives, Indicators, and Targets of PTSC, Arusha Project

S. No.	Objective	Indicators	Targets	Evaluation Results
1.	To train a cadre of postharvest trainers as "Master Postharvest Trainers"	Completion of a structured training program provided via e-learning	30 persons from 6 countries in 12 months	36 persons from 7 countries completed the program in 18 months
2.	To design, set up, and launch a PTSC in Kigali, Rwanda	Launch of a PTSC (open house)	Complete objective by 2012	PTSC was launched in October 2012 in Arusha, Tanzania
3.	To provide demonstrations and training programs, and conduct adaptive research	Provide postharvest training programs, research studies, and demonstrations at the PTSC	10 demonstrations 6 research studies 20 training programs in 1 year reaching 500 participants (250 men, 250 women)	10 demos set up 6 research studies completed 30 training programs reached 637 participants (230 men, 407 women) in 1 year

1. Training of postharvest trainers (including loss assessment and demo design)
2. On-site postharvest training and demonstrations
3. Adaptive research, including cost-benefit analyses of potential postharvest innovations
4. Postharvest retail shop (with tools, goods, and supplies) open to the public
5. Postharvest services for fees (e.g., grading, packing, storage, transport, and marketing advice)

18.7.3 DATA COLLECTION

Surveys, interviews, and observations were used to collect evaluation data. In order to investigate the outcomes as related to the project's theory of action, evaluators visited three key sites where Master Postharvest Trainers are currently working (in Ethiopia, Uganda, and Tanzania) and the site where the PTSC has been established (Arusha, Tanzania). We interviewed a variety of stakeholders, observed current practices, and characterized outcomes and results related to the five major components of the completed PTSC pilot project, along with general PTSC management. WFLO worked directly with local partners, farmer's organizations, women's groups, and local stakeholders, and hired local postharvest specialists, socioeconomists, and M&E consultants to assist with data collection.

Data on the 36 Master Postharvest Trainers were collected via a written survey distributed by email, since they were located in ten countries of SSA. Cluster (or area) sampling was used for collecting data from 50 local trainees in and near Arusha, since the trainee groups (who are farmers, food processors, and marketers) are geographically dispersed and it would have been practically difficult and very expensive to collect information from a randomly chosen selection of 50 individuals. Cluster sampling was achieved through a two-stage (or multistage) process: randomly choosing 10 trainee groups with which to work and then randomly selecting five individuals from within each of those groups. This multistage sampling approach helped to reduce the cost of quantification, since a large number of face-to-face surveys were done by local consultants, and the costs (for travel, making field visits, sorting, and organizing data) needed to be kept within the available budget.

18.7.4 EVALUATION RESULTS

The 18-month ToT resulted in 36 "postharvest specialists" who in turn trained more than 22,000 farmers, extension workers, food processors, and marketers in cost-effective practices and technologies that help them to reduce food losses and improve their incomes. The training occurred in seven countries in SSA over a period of 2 years. In addition, approximately 1,500 local farmers, traders, food processors, and marketers were trained via programs provided in Tanzania by the PTSC staff at the World Vegetable Center (formerly the AVRDC) site, Selian Agricultural Research Institute (SARI) staff at the Njiro site, and by locally and internationally based postharvest trainers. One hundred percent of those who have participated in training programs in Tanzania reported being satisfied with their experiences and making one or more changes in their practices that led to reduced losses of either fresh produce or processed products, as well as to increased earnings.

Best practices for postharvest extension were identified, and included providing a range of hands-on postharvest learning opportunities for clientele (men, women, and youths) from different communities and backgrounds, making sure that the trainees learned about how to calculate the costs and expected benefits of any technologies that were being promoted. Offering a range of options during training programs, rather than telling people what they should do, provided the best chance to learn about multiple options, try various practices, adapt what they have learned, and adopt only those practices that best suited their own current circumstances, (crops, postharvest problems), cultural beliefs, and budgets.

Because the training took place at the PTSC sites where many of these demonstrations have been established, the 50 evaluation respondents reported having seen the following demos, along with a rating of their usefulness (as most or least useful), shown in Table 18.3.

TABLE 18.3

Training Participant Reactions, Ratings, and Adoption of Demonstrated Improved Postharvest Practices

Postharvest Technology Demonstrations	# of 50 That Have Seen It	Rated the Demo as Most Useful	Rated the Demo as Least Useful	# of 50 That Have Been Using the New or Improved Practice
Use of shade	44	31		27
Gentle handling	33	24		17
Maturity indices	33	24		25
Improved containers	45	32		34
Sorting/grading	47	36		40
Hand washing/hygiene	45	10		35
Zero Energy Cool Chamber	45	33		20
CoolBot equipped cold room	37	3	24	0
Solar drying	44	34		29
Jam making	43	32	1	17

One hundred percent of those surveyed indicated that they were satisfied with their training experiences. Many of the respondents rated many of the demonstrations as "most useful." Only three of the respondents reported that they had not used one or more of the practices that they had first seen in a postharvest demonstration. The CoolBot equipped cold room was the most expensive postharvest technology offered as a demo, and was deemed the least useful by most of the respondents.

The three components intended to provide income for the PTSC (adaptive research projects, a postharvest retail shop, and postharvest services such as precooling and cold storage provided for local clients for set fees) were not fully implemented. While some of the planned adaptive research activities were initiated and provided supplementary funding for training at the PTSC, these research projects have not yet been completed or the results published. Partial success was achieved in the setup and stocking of a retail shop with an assortment of goods, supplies, and packaging materials, but very little inventory has yet to be replaced as the initial stock of goods have been sold. The planned "postharvest services to be offered to the public" such as packing, precooling, or cold storage were never offered by the PTSC, and the outcomes in terms of generating revenues for training and management of the PTSC from these three components did not meet the targets. The function of the PTSC in Arusha currently remains dependent on grants and support from the host organization.

Local stakeholders in Tanzania, however, have been participating and paying close attention to this project and then planning their own similar projects. They have already made improvements on the implementation of the five components as they adapted the model PTSC and then designed and launched their own projects. A variety of "Value addition centers," "Farmer services centers," "Postharvest training centers" and "Packinghouses/postharvest training venues" were developed in 16 districts. These were developed under programs implemented by the Ministry of Agriculture, Food Security and Cooperatives (including a packinghouse for 3,500 members of the cooperative "LUKOVEG" in Lushoto and a large citrus/mango packhouse/training center near Dar es Salaam for a farmer's association of 2,000 members), the Prime Minister's Office/African Development Bank Market Infrastructure, Value Addition and Rural Finance (MIVARF) project (with value addition centers in 16 districts) and Ministry of Agriculture/Tanzania Horticultural Association (TAHA) Farmer Services Centers (for three districts in southern Tanzania, one in Zanzibar, and a large packinghouse near the coast north of Dar es Salaam). Planning is underway for the LUKOVEG packinghouse to be duplicated with Ministry of Agriculture funding for farmers groups in two more locations near Lushoto.

The MIVARF project followed a similar design plan as the PTSC pilot project, starting with local needs assessment and commodity systems assessments, then working with stakeholders to select key crops and design appropriate training of trainers programs, select and procure equipment, and provide local training for farmers, food processors, and marketers. The MIVARF project has identified local "service providers" in each district who will serve as private sector partners. These were selected from among a pool of known successful local businesses, so MIVARF project managers believe they will be better able to operate the needed postharvest retail shops and provide maintenance, marketing support and other postharvest services for those trained at the postharvest training centers.

A recent request for applications (RFA) for a USAID project in Upper Egypt seeks to provide funding of up to US$23 million over 5 years, with a target of reaching 10,000 smallholder farmers to assist them with productivity, postharvest handling, and marketing (at a cost of US$2,300 per person). In comparison, international donor organizations such as the Bill and Melinda Gates Foundation recommend to applicants for funding that projects spend less than $240 per person in order for their large agricultural development projects to be considered a success. This is considered a difficult target to meet by most practitioners. The Horticulture Innovation Lab funded PTSC pilot project had a budget of less than US$500,000 over 3 years, and reached approximately 23,000 extension agents, small-scale farmers, food processors, and marketers in seven countries in SSA, at a cost of less than US$23 per person. As the PTSC model is adapted and adopted in more of the developing countries who could benefit from local capacity building in postharvest technology, it is hoped that these training successes can be replicated, and the results of the project evaluation can provide guidance on making improvements in the implementation of the components that were not a complete success.

18.8   M&E TRAINING AIDS/RESOURCES

- Using the Cost-Benefit worksheet:
  http://postharvest.org/PEF%20Training%20of%20Postharvest%20Trainers%20Manual%202016%20FINAL.pdf (pg. 36–40)
- Monitoring and evaluation of postharvest training projects:
  http://postharvest.org/PEFMonitoringandEvaluationofPHtrainingprojects.pdf
- Defining your mission, and setting goals and objectives:
  http://postharvest.org/PEF%20Mission%20goals%20objectives%20Postharvest%20MandE.pdf
- FAO e-learning website:
  http://www.fao.org/elearning/#/elc/en/courses/MEIA

18.9   CONCLUSION

There are many approaches to reducing food loss. Some may be more appropriate than others and some may work in some situations and not in others. In order to test and objectively evaluate the efficacy of a new practice or technology, it is necessary to measure and compare the result of the new approach to that of existing practices. This chapter describes some of the current models and methods of monitoring and evaluating projects and programs promoting PHL reduction practices. Accurate measurements are crucial to understanding barriers to adoption and calculating the return on investment of new technologies and practices. Using the included templates and worksheets can help assure fair comparisons of practices during postharvest project evaluation.

REFERENCES

Austrian Development Agency. (2009). *Guidelines for Project and Programme Evaluations*. Vienna, Austria: Austrian Development Cooperation. https://www.oecd.org/development/evaluation/dcdndep/47069197.pdf

Bennett, C. (1975). Up the Hierarchy. *Journal of Extension*, 13(2): 7–12.

Coote, H. C., Marsland, N. K., Wilson, I. M., Abeyasekera, S., and U. K. Klieth. (2002). Applied research and dissemination. Chapter 10 in Golob, P., Farrell, G., and J. E. Orchard (Eds.) *Crop Post-Harvest: Science and Technology. Volume 1 Principles and Practice*. Oxford, UK: Blackwell Science.

Donaldson, J. L. (2014). *Extension Program Planning, Evaluation & Accountability*. Nashville, TN: Tennessee State University. https://extension.tennessee.edu/eesd/Documents/StaffDevelopment/2014Orientation/NewEmployeeOrientation_August2014.pdf

IFRC. (2011). *Project/Programme Monitoring and Evaluation (M&E) Guide*. Geneva, Switzerland: International Federation of the Red Cross and Red Crescent Societies. http://www.ifrc.org/Global/Publications/monitoring/IFRC-ME-Guide-8-2011.pdf

Kitinoja, L. (2013). Returnable plastic crate (RPC) systems can reduce postharvest losses and improve earnings for fresh produce operations. White Paper No. 13-01. The Postharvest Education Foundation: La Pine, OR. pp. 26. http://postharvest.org/RPCs%20PEF%202013%20White%20paper%2013-01%20pdf%20final.pdf

Kitinoja, L. (2015). Final report for the TOPS evaluation project for WFLO. Alexandria, Egypt: World Food Logistics Organization.

Kitinoja, L. and D. M. Barrett. (2015). Extension of small-scale postharvest horticulture technologies—A model training and services center. *Agriculture*, 5: 441–455; doi:10.3390/agriculture5030441

Kitinoja, L. and H. A. Al Hassan. (2012). Identification of appropriate postharvest technologies for improving market access and incomes for small horticultural farmers in Sub-Saharan Africa and South Asia. Part 1: Postharvest losses and quality assessments. *Acta Hort* (IHC 2010) 934: 31–40.

LTS Africa. (2012). Evaluation of strengthening livelihoods and food security through reduction of grain port harvest and storage losses in Southern Somalia (IPHASIS) Final report. Nairobi, Kenya: LTS Africa.

NSF. (2002). *The 2002 User-Friendly Handbook for Project Evaluation*. Alexandria, VA: National Science Foundation. https://www.nsf.gov/pubs/2002/nsf02057/nsf02057.pdf

Patton, M. Q. (1991). Beyond evaluation myths. *Adult Learning*, 3(2): 9–10, 28.

Rossi, P. H., Lipsey, M. W., and H. E. Freeman. (2003). *Evaluation: A Systematic Approach*. Beverly Hills, CA: Sage Publications.

19 Summary and Conclusions

Majeed Mohammed and Vijay Yadav Tokala

The book provides details on the status of postharvest extension and capacity building activities in a wide range of regions (i.e., Sub-Saharan Africa, South Asia, Central Asia, Latin America, the United States, and Caribbean nations). The book content is divided into four sections. Section I explains postharvest loss (PHL) assessment methods. Food loss assessments provide concrete evidence on sources of food losses, exact point and amount of losses along the food supply chain, and assist in deciding suitable changes in handling practices or postharvest technologies that could reduce existing losses. Dr. Lisa Kitinoja described several new food loss assessment methods. These included the World Resources Institute (WRI) Global Food Loss and Waste Protocol, the United Nations Global Initiative on Food Loss and Waste Reduction (also known as SAVE FOOD) field case studies methodology (also known as the 4-S Approach), the International Food Policy Research Institute (IFPRI) technical platform on measurement and reduction of food loss and waste (which focuses on potential food loss and waste or PFLW), and an updated Commodity Systems Assessment Methodology (CSAM), led by the Inter-American Institute for Cooperation on Agriculture (IICA). Apart from exploring the historical background, objectives, and strengths in several field case studies involving food loss and waste (FLW), Dr. Kitinoja also explored notable gaps in the coverage of countries, crops, and food products, and weaknesses associated with each methodology. It was concluded that researchers and practitioners would benefit from the development and field testing of a hybrid methodology, incorporating the strengths and utilizing the best practices from each of the methodologies in current use, since high-quality FLW data is the best way to ensure that food loss reduction efforts will target the most important postharvest problems.

Chapter 2 provided recommendations that could be useful for future efforts to reduce gaps in postharvest research, training, extension, and capacity building in the Caribbean Community (CARICOM) region. This included researchers, extension workers, and value chain actors in training of trainer programs in postharvest loss (PHL) assessment. It was envisaged that a first-hand experience in observing the harvest, postharvest handling (PH), and measuring food losses along the supply chain from the farm to the markets is a practical way for stakeholders to learn about PHL and share their own knowledge and experiences. It was also argued that gathering of current data on PHL for key crops in more CARICOM countries is necessary, with a primary focus on the amounts and types of losses, the causes and sources of losses, the value of PHL, and the costs/benefits of any potential innovations or interventions intended to reduce PHL. By including researchers, extension workers, and value chain actors in these postharvest studies, local capacity will be increased.

Dr. Alpízar Ugalde discussed in concise terms the history and current improvements of the Commodity Systems Assessment Methodology (CSAM) to measure food losses. The useful nature of CSAM to technical specialists and decision-makers from ministries of agriculture, corporations, research institutes, and other institutions involved in making systematic improvements within agricultural value chains was explored in detail. The methodology systematically identified deficiencies throughout agricultural value chains that lead to food losses, and, at the same time, pinpointed solutions for improving their efficiency. CSAM divides the agricultural value chain into 26 components that are potentially important because the decisions or actions occurring at any point may affect the production, productivity, quality, or cost of the product at that point, or at some later point in the food system. It also presents tools for identifying and analyzing problems, proposing solutions and establishing priorities, with the aim of preparing project proposals and practical extension interventions. Since 2015, the IICA has led a process geared toward building the capacities of Latin American countries in relation to the CSAM methodology.

Section II gives a summary of different postharvest activities globally and covers a range of postharvest education, extension methods, and approaches being implemented around the world.

Chapter 4 provided an overview of the education/extension and capacity building activities being undertaken in the broad field of PHL assessment and loss reduction during the period of 2010–2017, and highlights achievements, and current and planned activities, in PHL assessment, training and extension, postharvest innovation systems, special events, and conferences. Nevertheless, the developing world continues to lack access to cold chain infrastructure during the PH and distribution of perishable foods. Despite significant enthusiasm amongst the development community, a gap in critically reviewing and synthesizing the abundant research on food losses still exists. It was discussed that current education and training on the measurements of PHL are plagued by the use of differing definitions, scopes, and ad hoc data collection methodologies. Effective education and training to promote capacity building at the national level will require increasing its scope to create a cadre of well-trained professionals. The aim is that these young postharvest specialists will lead their countries toward increased adoption and implementation of existing cost-effective postharvest interventions, while serving as PHL assessment team members, postharvest extension workers, private sector postharvest trainers, consultants on feasibility studies, and postharvest program/project evaluators.

Fonseca and Kitinoja reviewed the lessons learnt from several case studies that highlight key considerations with respect to integrating smallholders, gender relations, and nutrition in order to enhance extension/advisory services. Based on common factors found in countries studied, guidance was provided via the *Seven Steps Model*, with the following key objectives for more effective extension/advisory services:

- Meeting national development goals for food security, including access to adequate, safe, and nutritious foods
- Providing opportunities for enhancing rural livelihoods by creating new jobs in a growing postharvest/agro-industry sector
- Reducing food losses and waste in order to protect natural resources and to ensure that the national food system is being developed in an environmentally sustainable way
- Ensuring any underrepresented groups, such as women, smallholder producers, Small and Medium Enterprises (SMEs), traders, and youth involved in the food system have equitable access to productive resources and inputs, including agro-industry extension/advisory services.

Dr. Charles L. Wilson, the founder of the World Food Preservation Center® LLC (WFPC), presented the mission and the activities of the organization. The WFPC was established in response to a pending global food shortage that is destined to increase world hunger if not sustainably addressed. It was also articulated that the world's food supply is being shrunk by a rapidly exploding world population, deteriorating agricultural environment, and realities of global warming. Challenges of a global food shortage crisis and escalating world hunger could be alleviated by not only producing more food but by making every effort to reduce the 33% food losses that already exists. The WFPC was formed to address intellectual postharvest gaps in developing countries by (1) promoting the education (MS and PhD) of young student/scientists and secondary education students in developing countries; (2) having young student/scientists in developing countries conduct research on much-needed new postharvest technologies adaptable to their native countries; (3) organizing continent-wide postharvest congresses and exhibitions; (4) publishing much-needed new texts/reference books on postharvest technologies/methods for developing countries; and (5) developing a comprehensive database on all postharvest knowledge relative to developing countries with access portals for researchers, students, administrators, industry, businesses, and farmers.

The authors in Chapter 7 reiterated that integrated postharvest management of food is a major tool to combat the twin evils of loss and waste. They explained how the Amity International Centre for Post-Harvest Technology & Cold Chain Management (AICPHT & CCM), established in 2008 at Amity University, Uttar Pradesh, Noida, India has been actively involved in developing and applying various interventions to reduce PHL in fruits and vegetables through research, collaborative national and international projects, transfer of low cost technology to marginal farmers, and providing hands-on training to 1,745 primary beneficiaries. Major interventions included the use of portable shade, crate liners, cling film wrapping, low cost storage of fresh produce in zero energy cool chambers (ZECC) of 100 kg and 1 ton capacities, CoolBot fitted cool rooms, conversion of produce into value added products through minimal, semi-ready and ready to serve processing, solar drying, utilization of vegetable waste products for obtaining phytonutrients, and e-learning.

Dr Saneya Neshawy provided useful insights into the challenges related to policy-making in the postharvest technology of fresh fruits and vegetables in Egypt. The impact of poor PH practices, resulting in limited success in meeting international quality standards, was emphasized. Consequences of a broken supply/cold chains culminated in low-quality produce that was unsuitable for export and local marketing. Stakeholders in both the public and private sectors, along the production and supply chain, focused on quantity rather than quality. Women in rural regions play a key role in agriculture and therefore the demand for knowledge and information, training, and capacity building on postharvest techniques among women in Egypt was prioritized. Accordingly, an innovative approach named Gender Research in Arab Countries into information Communication Technology for Empowerment-Middle East and North Africa (GRACE-MENA) was initiated in Egyptian villages to assist women to become active managers of their own farms, by using information and communication technologies (ICTs) to train the women on postharvest technology, providing them with the information, and allowing them to explore more knowledge. Informative and updated webpage content on all the aspects of PH, including the methods to estimate the food losses, could be uploaded on the Internet for women to access the information whenever needed. Although the GRACE-MENA project has been concluded, a new book and a YouTube channel (https://www.youtube.com/channel/UCxyU5TE5FNRf-FvsUiE0ynA) are still in operation.

In Chapter 9, Bertha Mjawa discussed the investments in postharvest and food processing training, equipment, and value addition for reducing food losses in Tanzania. Food shortages due to prolonged periods of drought in several parts of Tanzania were also associated with the inherent weaknesses in the PH systems, which contributed to high food prices. Food losses also impacted on the environment, as land, water, nonrenewable resources such as fertilizer, and energy used to produce, handle, process, and transport food are eventually lost, since the products do not reach the final consumer. The author argued that mitigating PH losses can improve food security by increasing food availability, incomes, and nutrition, without the need to employ extra production resources, but approaches for PH loss mitigation in Tanzania have had little success. Smallholder farmers continued with subsistence post-production management practices. Adoption of PH interventions has often been poor because of innovation cost, sociocultural sensitivities, and inadequate knowledge among the majority of farming communities. Past interventions to reduce PH losses targeted improvement of handling and storage practices through the transfer of single and unconnected technologies, particularly for root crops and maize. The global consensus is that mitigating food losses, which occurs between harvesting and consumption, offered the single most enormous opportunity towards hunger alleviation. Basic skills should be harnessed for farmers to effectively manage PH losses, thereby empowering their knowledge capacities, as well as offsetting food deficits.

Section III included different innovative approaches to train the trainers, and to manage postharvest quality and reduce losses. E-learning is an effective way to reach practicing scientists and extension agents in distant places that may otherwise be difficult to reach. Chapter 10 explains the perspectives of The Postharvest Education Foundation (PEF) Global Postharvest E-Learning Program, a year-long mentor-guided training program in operation for the last 7 years, in which participants can learn at their own pace. In this program, each participant can select his/her own crops of interest and conduct assignments (readings, fieldwork, and written reports) on the topics most pertinent to their location. Most participants are university graduates, and many are working with national extension services, agricultural research centers, or private sector companies involved in the food supply chain. Successful trainees receives a Postharvest Train-the-Trainer Certificate of Completion and can purchase a Postharvest Tool Kit, valued at over US$400, at a subsidized price. PEF graduates are invited to travel to a closing workshop to work with the PEF team on implementing training programs for small-scale horticultural farmers. The graduates regularly communicate with the PEF and have had success in training and outreach. Their expertise in identifying the causes of food losses, their knowledge and use of postharvest technologies, such as solar drying, ZECC, processing, packaging, storage, and marketing, as well as postharvest training skills, have furthered their impact and their careers. PEF graduates also work as private consultants, while others have jobs as postharvest researchers, extension officers, trainers, and educators. All of these graduates have played an important role in reducing food losses in their communities.

Insight into the academic and outreach activities at The ADM Institute at the University of Illinois at Urbana-Champaign was highlighted in Chapter 11. The creation of the ADMI Village in Bihar, India as a testing ground to put PHL research into action was elaborated in detail, with emphasis on instructing local farmers in improved postharvest practices. The ADMI Grain Handling System successfully used grain dryers and hermetic storage bags in combination to improve the quality and quantity of grain that is harvested and safely stored. This encouraged the researchers and villagers to learn more about PHL interventions, the workable and affordable technologies, the awareness that is still needed at all levels about PHL prevention, and the training that will help the smallholder farmers learn firsthand that PHL is indeed a solvable problem. Technology centers in the village are conducted by trained lead farmers, who operate the dryers and other technology at the centers, thereby providing hands-on assistance to smallholder farmers. An ADM Institute-initiated bag subsidy program encouraged the farmers to purchase hermetic storage bags at a significant discount compared to market rates, thus making the bags accessible to smallholder farmers. Bihar partners were able to conduct several farmer training sessions on postharvest techniques, including some specifically for women, since women's participation dominated most of the postharvest processing work. The ADMI Village also provided a space for international conversations and education about PHL solutions.

Audiovisual media are some of the fastest and most effective ways of conveying a new practice or technology, and are understandable to the majority of the population, regardless of language spoken or geographic location of the observer. Dr. Barrett focused on the creation of postharvest and food processing videos for extension of postharvest technologies. YouTube videos were made to convey the principles behind three different topics: (1) PH of fruits and vegetables, (2) methods of evaluating the quality of both raw and processed fruits and vegetables, and (3) fruit and vegetable preservation methods. Videos with a short, simple format using overlaid text proved to be the easiest to translate and most likely to be understood by the largest variety of audiences. Short videos of less than six minutes were most popular. "How to" videos described how to use the various tools that a postharvest specialist would need to conduct evaluations. "How to Use a Color Chart to Increase Market Value" was the first video created. A "Technical Note" accompanied the video, and included the objectives, key concepts, materials required, background material, how learning will be reinforced/evaluated, references, and discussion questions. The second and third videos created were "How to Use a Refractometer" (https://www.youtube.com/watch?v=h1V7zqwMjrQ) and "How to Measure Temperature and Relative Humidity."(https://www.youtube.com/watch?v=8b3v0ky3OkQ) The videos and accompanying technical notes may be found at the same location as noted above. It was emphasized that videos are relatively easy to make and can be utilized to convey principles of postharvest technology, as well as to illustrate practical applications. Using the format of video with simple text overlays, videos can be easily translated into different languages for use around the world, using only a minimal amount of bandwidth.

In Chapter 12, key gaps along the production value chain were examined, where educational and/or technological opportunities exist for reducing PHL. Scientific Animations Without Borders (SAWBO) is one such research program, which investigated flexible, adaptive, and resilient educational solutions scalable to a diversity of value chain issues, including PHL. SAWBO involved creating and deploying linguistically localized animated videos that transfer scientifically grounded, best-practice knowledge to the widest possible audience (literate, low-literate, and non-literate alike) through the ever-widening world of digital infrastructure. Having investigated specific solutions to determine value chain gaps, SAWBO's research interests rest in issues of "knowledge chains," the scientifically flexible, adaptive, and resilient approach that is used to identify, solve, and deploy a given animated solution. SAWBO's research and responses to knowledge chains, which are analogous to and can be in support of value chains, as applied to problems of PHL, are discussed in three case studies. Animated videos on cell phones can fit into tightly constrained economic, social, and environmental spaces and times threatened by food insecurity and PHL, be it precarious markets, remote backwaters, gender roles, busy workdays, or resource-straitened circumstances.

In Chapter 13, Dr. Mohan gave a detailed description of the various postharvest tools (gadgets) developed as a professor at Tamil Nadu Agricultural University (TNAU). These include the probe trap, TNAU pitfall trap, and two-in-one model trap for monitoring and early detection of insects in stored food grains; indicator device and stack probe trap for population-level estimation; insect removal bin for farmers for early and sustainable removal of field carried-over adults and to prevent further buildup in storage; and TNAU ultraviolet light trap technology for insect control by trapping insects in warehouses. The insect egg removal device is also utilized for farm level/community level/traders level storage. Having found these tools to be effective, attempts were made by the author to transfer these tools/gadgets to end users in a systematic and scientific way, following all basic principles of agricultural extension. The author has used significant methods in the Transfer of Technology (TOT) effort: namely, developing entrepreneurs for producing an uninterrupted supply of gadgets to farmers/end users, conducting feedback studies, reaching out to the students both at school and college level by developing an educational kit and incorporating the techniques in the curriculum, and publication in journals and popular media. This has resulted in the significant spread of the storage loss reduction technologies to households, farmers, and government and private warehouses in India. In addition, the agricultural colleges, farmers training centers across the country, and school students in the Tamil Nadu state of India are using the TNAU education kit for teaching and training. In order to speed up the process of technology transfer to all users across India and also other developing countries, the process of attracting more entrepreneurs to produce and to market the products is still in progress as a future strategy for postharvest grain protection.

Section IV provides examples of postharvest extension and training activities being carried out in developing nations, and an overview of monitoring and evaluating practices intended to plan and implement postharvest food loss reduction projects.

Lessons learned from the establishment of The Postharvest Training and Services Center (PTSC) based at the World Vegetable Center's (WVC) offices in Arusha, Tanzania was discussed in Chapter 15. The PTSC was formed in 2012 with funding from the United States Agency for International Development (USAID) Horticulture Collaborative Research Program (HortCRSP, now known as the Horticulture Innovation Lab), led by Dr. Diane Barrett and Dr. Lisa Kitinoja, with technical guidance provided by the World Food Logistics Organization (WFLO). The goal of PTSC was to enhance access to the improved postharvest tools, services, and training to a wide range individuals and organizations involved in horticulture value chains. The center offers regular training sessions for extension agents, other agricultural trainers, and students from a wide range of countries in Sub-Saharan Africa. Training of farmers and small-scale processors has been undertaken as part of the linkage with farmer associations and their representatives. The trained representatives, in turn, disseminate the knowledge gained to other farmers and processors. Feedback from beneficiaries, particularly farmers, indicated that postharvest practices such as improving handling to reduce damage and grading of the produce are most widely adopted. More expensive interventions, such as the ZECC and solar driers, appeared to be of lesser interest to most farmers in Sub-Saharan Africa. More information on the costs and benefits of the investments is needed for these technologies to ascertain their value to farmers, traders, and commercial operators. The outreach and extension method that has been most effective was by "word of mouth," through the field days/visits to the PTSC by a wide audience (from extension agents to agricultural organizations and projects), as well as via radio and other media.

An evaluation of the USAID and University of Illinois Urbana-Champaign project, Farmer Advisory Services in Tajikistan (FAST), was discussed in Chapter 16. The project was successful in establishing an effective Extension and Advisory Services (EAS) in 11 USAID designated Feed the Future districts in Khatlon province in southern Tajikistan. The EAS model developed by FAST is suitable for scaling up, provided that some cost-saving modifications are included. The EAS has been developed for household and small commercial farming, with the aim to produce enough nutritious food for the household family to consume, and to sell some surplus. The project faced the challenge of recruitment of suitable female field staff to work with, and the availability of short-term technical assistance. The project provided considerable resources for intensive training and coaching of key staff members, to bring them up to par with the required participatory attitudes with modern technical production skills. Through learning by doing, the project had to find out the positions and the minimum numbers of staff required to implement the EAS. FAST has the potential to develop an agricultural extension model to reach a larger number of small commercial and rural households in Tajikistan, and to contribute towards rural livelihood improvement, household food security, and empowering rural women.

Professor Mahajan has described the significance of postharvest horticulture and activities of the Punjab Horticulture Postharvest Technology Centre (PHPTC), Ludhiana, Punjab that was established in 1998. The PHPTC focused on conducting training programs in postharvest management and marketing of horticultural crops. This center has been working as a nodal agency to guide farmers, entrepreneurs, and government agencies to establish cold chain infrastructure and to provide technical inputs to all the stakeholders for the safe PH of fruits and vegetables for domestic and export marketing.

Kitinoja and Zagory described some of the current models and methods of monitoring and evaluating projects and programs promoting PHL reduction practices. Emphasis was made on the significance of taking accurate measurements in order to understand barriers to adoption and calculation of the return on investment of new technologies and practices. Templates and worksheets were recommended to assure fair comparisons of practices during postharvest project evaluations. The chapter provided an example of a postharvest project evaluation, training aids, illustrations, and workshop presentations.

CONCLUSIONS

Different postharvest food loss and waste assessment methodologies are being implemented by various workers around the globe, which include both sampling and ad hoc survey methods. But due to lack of proper standard methodology, to date, the quality of the available information is not good enough to allow systematic identification of appropriate solutions to reduce losses. The data available do not provide clarity on adoption of cost-effective changes in the existing practices or for profitable postharvest agribusiness investments. Establishment of PHL assessment methods applicable globally with relatively uniform definitions and capacity to provide high-quality data is the immediate need to reduce PHL.

Several regional and global outreach projects are being implemented to train different stakeholders of food value chains to reduce PHL and to build essential postharvest capacity. Encouraging effective education and training from secondary school to university level in order to promote capacity building at country level will enhance the scope to create a cadre of postharvest specialists.

These postharvest specialists will be capable to lead their countries towards encouraging cost-effective postharvest innovations, serve as postharvest trainers, consultants on feasibility studies, and effective project implementers and evaluators. Different approaches, including e-learning, videos, animations, and many others, are presently available to train a diverse range of people, globally, towards PHL reduction. Effective postharvest sensitization through these educational technologies will play a major role in reducing food losses.

Postharvest extension platforms concerned with capacity building should be a one-stop location for all the postharvest goods, services, training, and consultancy and should be well-equipped with the essential postharvest infrastructure to effectively deliver key technical information to individuals of different backgrounds, needs, and interests.

Developing a standardized food loss assessment methodology and effective postharvest capacity building through innovative training methods and extension platforms will enable the world to achieve the United Nations Sustainable Development Goal (SDG) target 12.3, to reduce worldwide food losses and waste by 50% along the production and supply chains by 2030.

Index

Note: Page numbers in italic and bold refer to figures and tables respectively.